Engineering Design in the Multi-Discipline Era

A Systems Approach

Engineering Design in the Multi-Discipline Era

A Systems Approach

by

Paul R Wiese

and

Philip John

Professional Engineering Publishing Limited
London and Bury St Edmunds, UK

First published in 2003 by Professional Engineering Publishing Limited, UK.

ISBN 1 86058 347 4

A CIP catalogue record for this book is available from the British Library.

Printed and bound in Great Britain
by Antony Rowe Limited, Chippenham,
Wiltshire, UK.

Related Titles

Book Title	Author(s)	ISBN
Advances in Vehicle Design	J Fenton	1 86058 181 1
Aircraft Conceptual Design Synthesis	D Howe	1 86058 301 6
Constraint Aided Conceptual Design – Engineering Research Series (ERS9)	B O'Sullivan	1 86058 335 0
Design for Excellence Engineering Design Conference 2000	S. Sivaloganathan and P T J Andrews	1 86058 259 1
Design and Manufacture for Sustainable Development	B Hon	1 86058 396 2
Improving Maintainability and Reliability through Design	G Thompson	1 86058 135 8
Managing Engineering Knowledge MOKA: Methodology for Knowledge Based Engineering Applications for the Moka Consortium	M Stokes	1 86058 295 8
Sharing Experience in Engineering Design – SEED 2002	M A C Evatt and E K Brodhurst	1 86058 397 0

For the full range of titles published by Professional Engineering Publishing contact:

Marketing Department
Professional Engineering Publishing Limited
Northgate Avenue
Bury St Edmunds
Suffolk UK
IP32 6BW
Tel: +44 (0) 1284 763 277
Fax: +44 (0) 1284 704 006
email: marketing@pepublishing.com
Online bookshop at www.pepublishing.com

About the Authors

Professor Paul Wiese is the Royal Academy of Engineering Visiting Professor in the principles of engineering design at both the Open University and at Cranfield University. He studied at Liverpool University where he obtained a first-class degree in mechanical engineering. His background is in industry where he held senior executive positions in the both the Engineering and the Manufacturing Departments at Rolls Royce, Bristol, including the posts of Chief Production Engineer and then Chief Engineer for the Olympus engines in Concorde. He was Technical Director at Molins plc for seven years and a Vice President of the Langston Corporation in America. At Molins he headed the company's involvement in the UK's national research programme into the design of high-speed machinery, chairing the Programme Management Committee of the EPSRC/DTI LINK research programme of the same name for many years. He was later with Ricardo Aerospace, the Consulting Engineers. Paul has extensive industrial experience in the design of mechatronic machines and has written numerous articles on mechatronics and intelligent control. He contributed the section on Design in the Open University's course 'Mechatronics: the Design of Intelligent Machines'. Currently Paul is active in the research into the intelligent geometry compressor, a concept he pioneered. This is his first book.

Professor Philip John holds the Chair in Systems Engineering at Cranfield University. He studied at Imperial College, London, doing a PhD in Mechanical Engineering, before spending 18 years in the defence industry. His industrial career encompassed a wide range of roles covering the whole life-cycle, from operational assessment and research, through concept and feasibility studies and full-scale development and in-service support. Over the years, Phil had systems engineering, project management, and functional management responsibilities, including leading multi-disciplinary integrated development teams, systems engineering research, and major business process re-engineering programmes. His experience covers both UK and multi-national defence programmes and, before moving to academia, he was Head of Systems Engineering for a multi-national prime contractor organization, covering the whole scope of systems engineering, such as requirements engineering, system design, ILS, ARM, human factors, safety, systems proving, and simulation and modelling. Phil joined Cranfield in 1999 as the Professor of Systems Engineering and leads the growing Systems Engineering Group. He is a member of several national bodies, including two national advisory committees (for systems engineering and for synthetic environments), the UK MOD's Requirements Capability Working Party, a Defence Scientific Advisory Committee on Systems and the IEE's Executive Team on Systems Engineering. Phil's research interests include the development of complex integrated systems, risk management, and a whole system-through-life approach.

Contents

Acknowledgements **xiii**

Frontispiece **xv**

Foreword **xvii**

Chapter 1 The Role of Engineering **1**

1.1 Introduction 1
1.2 Early engineering and the need for compromise 4
1.3 Engineering integrated machines 5
1.4 The changing demands on engineering products 6
1.5 Market drivers 6
1.6 Mass customization 7
1.7 Systems integration 9
1.8 What this book is about 10
1.9 Who this book is for 11
1.10 References 11

Chapter 2 The Evolution of Engineering Design **13**

2.1 Introduction 13
2.2 Early design evolution 15
2.3 Case study: Simple sub-system integration 18
2.4 Case study: Vehicle ignition systems 21
2.5 Modern technologies 23
2.5.1 Control 24
2.5.2 Sensors and actuators 25
2.5.3 Power supplies 26
2.5.4 Controlled drives and mechatronics 27
2.6 Case study: Force versus timing 30
2.7 References 32

Chapter 3 The Need for a Systems Engineering Approach to Engineering Design 35

3.1 Introduction 35
3.2 A traditional view of engineering design 35
3.3 A traditional view of systems engineering 38
3.4 Systems and systems influences in engineering design 41
3.5 Dealing successfully with systems 44
3.6 Achieving a systems engineering approach in engineering design 45
3.6.1 Building a systems understanding 46
3.6.2 A holistic approach 47
3.6.3 Evaluating decisions in systems terms 48
3.7 Implications for the engineering profession 48
3.7.1 Education 49
3.8 Summary 50
3.9 References 50

Chapter 4 The Enterprise Environment for Engineering Design 51

4.1 Introduction 51
4.2 Early engineering design 52
4.3 Formalized (system) engineering design 54
4.4 Enterprise expectations 56
4.5 The nature of engineering design in modern enterprises 59
4.6 Strategic design 61
4.7 Simulation 62
4.8 Summary 65

Chapter 5 Control and Embedded Intelligence 67

5.1 The traditional view of control 67
5.2 Control as a more flexible system 68
5.3 The perception–cognition–actuation model 69
5.3.1 Sensor fusion 70
5.4 The need for embedded computing and communications 71
5.5 The nature of embedded computing and the concept of artificial intelligence 71
5.6 Uncertainty and intelligence related to games and industrial processes 72
5.6.1 Noughts and crosses 72

5.6.2 Chess 72
5.6.3 Poker 72
5.6.4 Robot football 73
5.6.5 Industrial processes 73
5.7 Techniques of AI and how they are applied 74
5.7.1 Vision system for cheque reading 74
5.7.2 Fuzzy control for central heating control 74
5.7.3 Automatic lawn mower 74
5.7.4 Intelligent leg-joint prosthesis 75
5.8 Distributed intelligence using autonomous intelligent agents and the concept of emergent behaviour 75
5.8.1 Agents 75
5.8.2 Autonomous intelligent agents 75
5.8.3 Communities of agents and emergent properties 76
5.8.4 The concept of proliferating simple autonomous agents to propagate emergent behaviour 77
5.9 Current technologies and their significance 78
5.9.1 Neural networks 78
5.9.2 Image capture and interpretation for machine vision 79
5.9.3 Searching techniques 79
5.9.4 Deterministic methods 80
5.9.5 Probabilistic methods 81
5.10 Rule-based systems for diagnosis and decision-making 82
5.11 Fuzzy logic for control purposes 82
5.12 How do these concepts fit into systems engineering? 83
5.13 Reference 83

Chapter 6 The Challenge of Idea Sourcing 85
6.1 Competing in a knowledge-based world – the challenge of idea sourcing 85
6.2 Mechanisms that can assist technology awareness and transfer 90
6.3 In conclusion 92

Chapter 7 Case Studies 95
7.1 The wrapping machine 95
7.2 The goods distribution system 101
7.3 The local bus network 102
7.4 Manufacturing systems and development requirements 103

7.5 The missing audit trail 105
7.6 Designer knows best 105
7.7 Electric brakes 106
7.8 Agriculture – farming as a system 106
7.8.1 The automatic milking machine 106
7.8.2 The mini-tractor concept and use of satellite images 107
7.9 The affect of accountancy procedures on engineering decisions 107
7.10 New technologies associated with systems engineering 108
7.11 Examples of technology mix in industrial sectors 108
7.11.1 Aerospace 108
7.11.2 Agriculture 109
7.11.3 Automobile 109
7.11.4 Gas turbines and rotating machines 109
7.11.5 Manufacturing 109
7.11.6 Medical engineering 109
7.11.7 Modularity 109
7.11.8 Process industries 109
7.11.9 Transportation 110
7.11.10 Unsuccessful systems 110

Chapter 8 Final Remarks 111

Index 113

Acknowledgements

This book could not have been written without help from others and we would like to particularly thank:

Dr Anthony Lucas-Smith, Senior Lecturer at The Open University, who contributed Chapter 5 on Control and Embedded Intelligence.

Professor Chris Pearce, Technical Director of INBIS Group plc, based in Bristol, England who agreed that we could include as Chapter 6 his Keynote Address to the Royal Academy of Engineering Visiting Professors Conference in April 2002.

Frontispiece

'And it ought to be remembered that there is nothing more difficult to take in hand, more perilous to conduct and more uncertain of success, than to take the lead in the introduction of a new order of opportunities. Innovation has many enemies; all those that have done well under the old conditions are lukewarm defenders of those who may do well under the new. This coolness arises partly from fear and partly from the incredulity of men who do not believe in new things until they have long experience of them.'

Anon.

Foreword

This book is about relationships. Relationships between technologies, engineering disciplines, enterprises and, above all, people. In the modern world, we all experience a web of relationships and these are becoming stronger, more wide-ranging, and increasingly complex. In short, we live and work in an increasingly joined-up world and nowhere is this more apparent and important than in the realm of engineering and technology. Indeed, it is the dramatic developments in technology over the last century that have been the driving forces behind these changes. This book explores the implications of these changes for the way in which we should think about technologies and conduct engineering design – the crucial decisions that are made, the expectations they must satisfy, and the relationships between the engineering domains that are involved. The need to address the nature of engineering design in the modern world is widely recognized, as shown by the increasing focus on, and activity in, mechatronics and systems engineering – and these again emphasize the importance of understanding and dealing with relationships.

Our original intention for this book was to set out a range of case studies, chosen to cover a wide range of technologies and applications and so to provide a resource for discussion and education in engineering design. However, in trying to explain the nature of the case studies and the messages to be drawn from them, we realized that there is a need to set out the context for engineering design itself in order to stimulate debate on the crucial associated relationships. For example, when viewed at a particular point in time, how does engineering design relate to technological limitations? How does it relate to the enterprise expectations? How do different engineering domains relate to each other within processes and organizations? These (among others) are crucial questions that must be addressed. We must discuss how today's increasingly joined-up world affects these issues, and the implications – the way in which we should conduct engineering design when faced with increasingly joined-up, wide-ranging decisions.

We still make extensive use of case studies throughout the book, and still hope that these provide a useful resource in themselves, but we also use them to discuss the very nature of engineering design and the changes it faces in coping with the modern world.

The book is aimed at all those who come into contact with the impact of modern technologies, from those involved in engineering design itself to the leaders within engineering enterprises who seek the effective exploitation of modern technology, and from students to experienced engineering specialists. It is intended to stimulate debate about the nature of the challenges we face in today's engineering world and on the decision-making and relationships in the modern world. We would encourage all to join in the debate.

Chapter 1

The Role of Engineering

1.1 Introduction

Engineering is all about applying scientific principles to the design and manufacture of useful items. The Institution of Mechanical Engineers has a saying: 'Nothing moves without mechanical engineers!' While the field is much wider than just moving objects, the inference is clear, yet many of the best-known engineering artefacts have reached their present state of design by a very long period of evolution. George Stephenson (Fig. 1.1) could dismantle a modern car gearbox and not be unduly surprised by what he would find inside. Yet today, we suddenly find ourselves in a totally new and exciting environment, with multi-discipline skills being necessary for

Fig. 1.1 George Stephenson

design, and many traditional design approaches no longer adequate for modern conditions.

Our aim in this book is to look at present-day design and to demonstrate how a systematic multi-discipline design can now provide a performance unobtainable by traditional design techniques, and how this approach must be integrated into a much wider overall system. This is what 'systems engineering' should be, but the term already has definite connotations in certain fields (such as electrical engineering). You may find some of our definitions new to you, but we hope we can demonstrate the need for an integrated look at all aspects in the design stage. The key to this is of course the power of the computer, now coupled with electrical and mechanical forms of actuation that can achieve new levels of performance through intelligent control. Perhaps some examples will make this clearer.

In many reciprocating engines, as well as in high-speed processing machines, particular displacement, velocity, and acceleration profiles are required at discrete positions in the operating cycle. For instance, the opening and closing of the inlet and exhaust valves in an IC engine are timed necessarily at precise points in the rotational cycle, as shown in a typical timing diagram (Fig. 1.2). In a processing machine, the timing and displacement of a product 'pusher' need to happen at a precise point in the operating cycle (Fig. 1.3). These position and velocity profiles are usually

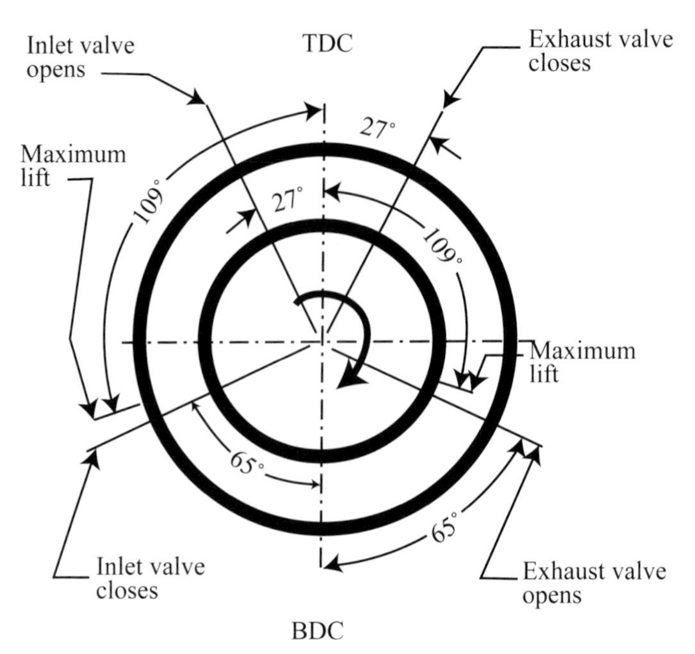

Fig. 1.2 IC timing diagram

obtained by very complex mechanical systems, using various cams, gears, linkages, chain drives, Geneva mechanisms, and the like. The result is that the design of the transmission system can dominate the whole machine design. Today we can generate almost any position and velocity profile we choose through the use of software and special computer-controlled electric motors and actuators, using intelligent control. The design of machines has thus been liberated from the constraints of the transmission system and design can now be concentrated on the needs of the process itself.

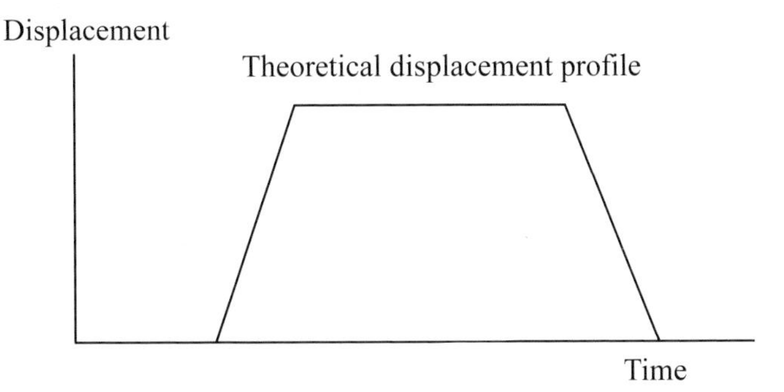

Fig. 1.3 Timing diagram wrapping m/c

In the examples quoted, the valve timing in an IC engine can be made variable to suit different load and speed requirements. Would we then still use a poppet valve/cam mechanism? In the processing machines, the product can now be placed precisely with local actuators, obviating the need for complex linked mechanical systems. Design has thus become, of necessity, a multi-discipline systems approach to the desired solution, where these new techniques allow us to carry out a critical reappraisal of the real needs of the design, rather than the compromises forced on us by single-discipline solutions. In the examples quoted, the timing can be chosen to achieve the exact needs of the process, rather than a compromise dependent on the limitations of mechanical transmission systems. We can now look, therefore, at applying full scientific principles to engineering design, looking at the exact needs of the system and having fewer constraints on our design approach, whether we are designing machines or buildings, microprocessors or aircraft. This is the important difference offered by a systems engineering approach to design.

A more current example of this approach is in the ABS braking system now fitted as standard to many cars. The technique of improving braking performance by automatically releasing the brakes as soon as a wheel locks (and skids) is not new. However, the incorporation of sensors and intelligent control in ABS systems has enabled the technique to be developed into a new sub-system that can form part of an integrated engineering approach to the whole vehicle design, including here both braking and steering functions. Compare this with the previous mechanical braking system, which had little affinity to the other vehicle sub-systems.

1.2 Early engineering and the need for compromise

As scientific principles were first discovered, uses and applications were quickly found to exploit these new discoveries to the full. Just when we would call this progress 'engineering' is hard to define, but we can certainly accept that at the start of the Industrial Revolution, engineering principles were being applied. The early engineering pioneers quickly found that compromises were necessary, not only due to the knowledge available at that time, but also because of the type of materials that had to be used. The surprising thing is the amount of this empirical thinking that established the guidelines that we still use today, and which we are only now re-examining in detail.

It is interesting to examine the evolutionary trail, and we can see how simple machines or sub-systems were eventually combined to produce more complex machines. This process went on unabated for over 200 years, and is still in vogue today. The steam engine developed into the steam railway engine by the application of a stationary engine to a moving platform (**1**). George Stephenson developed the idea further and laid the foundation for the later generations of locomotives. Similarly the car was derived from the application of a stationary gas engine to a horse-drawn cart. These machines fulfilled an overall purpose that could not be achieved by any one sub-system on its own, and so began simple systems technologies. In these simple systems each sub-system has a clear role to perform, such as braking, steering, etc., and the overall system is constructed by assembling the sub-systems to work as one unit. The interactions between these sub-systems are simple and dominated by the interfaces between them – mechanical interfaces in early machines, plus electrical in later machines. The sub-systems are therefore themselves dominated by a single technology and thus by a single engineering specialization. It is this aspect which is now changing. At the same time, machines and systems are increasing in complexity, with a larger number of components, higher operating speeds, and demands for ever-greater quality in build, operation, and output. In themselves, the machines still remain fundamentally the same, and in systems terms they are still 'simple', the sub-systems being predominantly linked to a distinct contribution to the overall system, and the relationships between sub-systems being dominated by their interfaces yet with clear allocation of requirements.

1.3 Engineering integrated machines

If we look at the way some technologies have developed that are changing the nature of these simple systems, we find information and communications technologies change the nature of the interactions between sub-systems, and provide new methods of linking them in very different ways. Combining these with other developments (reduction in size, increase in power-to-weight ratio, etc., these changes enable very different systems to be designed, leading to more opportunity, more innovation, and the need for more sophisticated design decisions. In particular, systems are no longer dominated by the performance of their sub-systems and their interfaces, but by genuinely emergent properties, i.e. complex systems behaviour. This in turn changes the relationships between the engineering disciplines, and it is this opportunity which must be grasped in design. The new technologies are producing a definite change in the rate of design progress, and they present difficulties in working and integration, as well as new opportunities.

Progress with the development of sensors and actuation also enables integrated systems to be designed. Although a varied range of sensor and actuator types has been available in one form or another for many years, recent developments have considerably widened the scope and use of these components. In particular, miniaturization has enhanced the use of sensors in many fields and actuation is the key to many modern technologies. The Japanese, for instance, have been active in designing medical actuators that can be miniaturized to the point where they can be introduced into a person's veins, to carry out medical operations from inside the human body. Again, much of the work that has been carried out on software-controlled electric drives is directly applicable in the field of actuation, while significant progress has been possible with hydraulic and pneumatic methods of actuation. These are all significant advances in the enabling technologies that allow us to consider our approach to design in a more robust way.

Later chapters look at these advances in a more detailed way and look at the forward view, particularly in control and embedded intelligence, giving us a strategic perspective (rather than the more usual tactical approach) to design for the future. It should be emphasized that these developments apply across engineering, and should not be seen as solely in the realms of mechanical and electrical design. We should not forget that the Japanese built a bank in Nagoya in the early 1990s, using automatic mechatronic systems.

1.4 The changing demands on engineering products

Just as the relationships between engineering disciplines are changing, so too is the world outside and with it the demands it places on engineered products. Mass customization, bespoke production, 'build to order not to stock', life cycle requirements, environmental pressures, and health and safety aspects are all demanding new approaches to the output of the overall engineering design system, while at the same time the rate of change itself is ever increasing.

The implications are that, while system designs are more highly integrated, engineers can no longer afford to work in isolated, single-discipline teams. Design solutions have to be multi-discipline to achieve even satisfactory results let alone competitive solutions. Inevitably, this need will have an influence on future teaching in engineering, and some establishments have already recognized this need, and have or are planning courses in mechatronics.

1.5 Market drivers

Today's markets are changing, and changing rapidly. To be competitive there is a need to offer variety in products, coupled with competitive prices. Remembering the old adage that the selling price is what the market will bear – not the manufacturing cost with a profit margin added – the requirement in all fields is flexibility for sales appeal and flexibility in production, to keep costs down.

We are used to basically three forms of manufacture: mass production, batch production, and bespoke manufacture. In the past, machine design, shop layout, shop control, and management hierarchies have all been based on these three forms of manufacture, with each type using slightly different strategies to succeed. The current and increasing market demand for bespoke articles at mass-production speeds and prices is putting our existing manufacturing strategies under great strain, and the use of inflexible machines compounds these problems. The new market requirement is known as 'mass customization', and there are already numerous examples in the outside world to define this new requirement. This trend can be seen in products as diverse as jeans and bicycles, while it has been available in a muted form in motor vehicles for some time. From a design point of view, the clear indication is the need for change to greater flexibility and speed in the manufacturing process, a need that cannot be achieved with the dedicated lines typical of mass production used for so many years. It is the exact antithesis of Henry Ford's 'You can have any colour as long as it is black'.

It is important to realize the importance of these changing market conditions. Like most things in a competitive market, the orders and the profit go to those who respond the fastest. However, traditional design approaches are not best suited to making versatile machines, and cannot easily cope with remote programming, integrated systems requirements, and the like. Can simple mechanical sub-system integration suffice? Possibly it can in a purely 'Can we achieve it?' mode, but very definitely not in a competitive environment. Again, increased flexibility requires changes in all the modes of manufacture mentioned above, and the difficulties in changing management structures and attitudes can be just as difficult as changing engineering design practices. At the same time, demands for flexibility can be partially met by flexibility in the products to be manufactured, and it is inherent in systems engineering that this particular trade-off has to be done. Modularity is a concept that has much to offer on both sides of this particular equation.

In areas away from engineering, increased flexibility is also demanded, as the case study on the goods distribution system shows (Section 7.2). These requirements, usually known as 'technology pull', are particularly strong at the present time with the current demand for ever-increasing competitiveness. As with most things in life, these demands eventually call for changes or improvements in engineering products, so strategic design philosophies need to be inherent in our systems thinking.

1.6 Mass customization

Mass customization is about being responsive to customers' requirements and customers' needs. It is a paradigm shift that is partly with us already and one that will expand rapidly as customers realize the potential of having exactly what they want, rather than whatever is closest to their needs. It is bespoke products at mass-production speeds and prices, and can be applied across all fields; but it also demands complete flexibility, allied to producing the products involved. It is closely linked to agility, although agility is the wider function since it is applied to many other spheres and products apart from mass customization. However, the main drivers are the same and the same approaches to design are needed.

A case study may illustrate mass customization better. For years, bicycles have been mass-produced, based on set frame sizes and angles. While not quite 'one size fits all', the approach has been to standardize on a small number of frame sizes, and to persuade customers which 'standard' size is best fitted to their needs. Racing cycles are a different matter, but that market is specialized. A Japanese manufacturer has developed a technique that enables them to make individual cycle frames to the exact individual customer's needs (as regards frame size and frame angles). They have also developed this manufacturing system with sufficient agility that they can supply these individual bespoke cycles for next-day delivery, at standard prices. An interesting side effect of this is that their customers do not believe that their cycle can be manufactured this quickly if it is truly to their individual sizes, so the delivery of the cycles has to be delayed by a couple of days to retain credibility! In clothing, bespoke tailoring is a long-established tradition, but again mass customization is bringing bespoke tailoring to such areas of mass-produced garments as ladies' jeans. One manufacturer has recently introduced their individually sized 'made to measure' jeans based on their mass-production flexibility with use of a much increased range of panel sizes. The key to these new practices is the close link between the end customer and manufacturer, much heralded in the larger companies with very-high-cost products (such as aircraft) but now being applied right down to purchases by individual members of the public. As has been mentioned, it can be confidently predicted that once customers are offered exactly what they want, rather than something that comes closest to their needs, there will be no turning back. For the customer it can be seen as a return to the days of 'made to order', but now at mass-production speeds and prices.

The introduction of mass production was a paradigm shift that brought with it the mass marketing, mass distribution of goods, mass media, etc., with which we are all so familiar today. These powerful 'technology pull' factors have been breaking down as suppliers look for better ways to market their goods, and the introduction of a choice of features (and therefore individuality) is seen as one way to go. Mass customization is really the end product of this trend. While recent technical developments have provided the 'technology push' required to provide the agility required for mass customization in the mass market place, two things stand out. Mass customization is predominantly a rapidly accelerating technology pull regime and it will require effective systems engineering in all its aspects to make it happen.

Some of the techniques required to achieve mass customization do not impinge on design in a direct way, but they do give an indication of the width of view necessary, in a systems engineering sense. If individual orders are to be produced quickly, it is essential that undue time is not taken in bringing materials to production machines. We may design amazing speed and agility into the machine, but it will not produce any products if it is not fed with the required materials at the right time! This may seem very obvious, but the writer was involved some years ago at a large batch-production factory in an exercise to increase cutting speeds. A big exercise was carried out, looking at feeds and speeds on metal-working machinery, and significant improvements were obtained – well above those known through national standards. Although spectacular, and it has to be said politically very important, a separate exercise also identified that the flow of the materials through the workshops was the real bottleneck and accounted for over 70 per cent of the lost time. The improvements in feeds and speeds contributed about 12 per cent in the overall picture. So systems engineering needs to look at all influences, which until today have largely remained in 'stovepipe' sectors of thought, industry, and engineering for so very long. These changes entail embracing the use of modern IT techniques (among others), but the mass proliferation of personal IT communications has prepared people for this impact and the need for it. The mental preparation that needs to precede change is already present, so the need for change can be readily identified. Can commerce, industry, marketing, and the media cope with such a change? Perhaps many of our well-known institutions are past their own 'best by' date! It is worrying how long-established and historical patterns of behaviour are so difficult to change. Can the bureaucratic organizations of the past and of today also have the agility to cope with mass customization?

1.7 Systems integration

We shall see an early example of systems integration in the case study of the horseless carriage in the next chapter. However, the current need for complex systems and the purchase of constituent parts of the systems from separate suppliers (and in some cases separate nations) has brought with it the need for perfect integration at the interfaces. While this is a most powerful requirement, it has rather led to the belief that if the interfaces can be controlled adequately, a complete system can be built from sub-systems. This completely misses the point of how much more can be achieved by correct systems engineering, rather than treating a system like a Lego set that can be built successfully by assembling predetermined building blocks. The systems integration approach is much seen in military thinking, and certainly there are enough horrendous examples of incorrect interfaces from WW2 to drive this message home very forcibly. We must not confuse systems engineering with systems integration, however – integration is a sub-set of systems engineering. We can also see that to design properly in a systems engineering sense will need a paradigm shift in many existing industrial design offices. Sub-system design with good system integration has served us well for very many years, and will continue to do so. How else would we have achieved complex systems like fighter aircraft with different sub-systems designed and built in different countries, in different companies, with different cultures and standards?

What this all illustrates is the need for very clear thinking and very clear interpretation of the total requirements right at the beginning of the exercise – the real start of systems engineering. For instance, with a modern car, how do we balance life-cycle costs, disposability, reliability (both short term and long term), etc., against the latest gizmo to attract buyers? Should the needs of the third, fourth, and fifth owner on be considered? Will our intelligent-control philosophies and hardware be adequate for a 12-year life?

The answer to many of these questions is standardization, but the demands of mass customization are driving us in the opposite direction. The old design philosophies of simple sub-systems and system integration do begin to look simple by comparison!

1.8 What this book is about

This book is intended to set out the scope and techniques of current design of multi-discipline systems and the management of this design effort. It explores the trends from simple systems (built up with simple sub-systems) to highly integrated systems, and considers the implications of this on the design processes, methods, and tools, and on all of the system lifetime issues. We show that many past design principles and practices must be revisited, or at the very least applied in a new context, in order to cope with these new system challenges and to grasp the new system opportunities. The approach we advocate involves all engineering disciplines, and the whole 'system' here includes the designer, the customer, the production methods, and indeed the organizational needs as well. It covers the changing nature of engineering, the evolution of a systems approach, and the implications of modern technologies such as intelligent control, and gives examples through extensive case studies to underline the basic thinking that is necessary with this approach. At the same time we look at design techniques such as strategic design, systems integration, and reconfigurable systems, and examine current techniques and their applicability in the new environment. Case studies are a powerful way to illustrate principles, and the examples chosen should make the reader think through the underlying reasons behind particular approaches. All the case studies are new. However, not all of them have succinct answers! Some will make you think of possible ways forward and are intended to ask you to form your own opinions.

1.9 Who this book is for

The book is aimed at a wide range of readers, virtually everyone who is involved in design in the widest sense. We hope we show that as different technologies develop, design becomes a truly multi-disciplined process, necessitating knowledge of not only different engineering disciplines, but also awareness of the wider forces acting on the designer and the design process. The subject covered by this book should appeal to anyone directly involved in design, of whatever discipline, particularly those experiencing some of the new pressures that the changes we describe are bringing. Those involved in system integration should have their horizons broadened, and students can be prepared for the world as it is and as it is becoming. Project leaders should see or confirm a new dimension, while administrators and people involved with political decisions can get a feel for the changing scene, and hopefully be able to input some of the changes that are needed away from just engineering. We hope a wide audience can benefit by having the changes more clearly laid bare, since so much of systems engineering seems to be a mere amplification of systems integration.

One point we have tried to highlight is the historical development of human skills to control the systems we have built in the past, and how artificial intelligence has already replaced these skills in some sectors and will do so even more in the future. As we have said, the case studies are new; we hope they are readable and of interest in a fundamental way.

Keys:

- Systems engineering, not just sub-systems integration.
- Multi-discipline design.
- Evolution no longer sufficient.
- 'Technology pull' through greater customer expectations.

1.10 Reference

(1) **Briggs, A.** (1982) *The Power of Steam*, Chapter 4, Bison Books, University of Nebraska Press.

Chapter 2

The Evolution of Engineering Design

2.1 Introduction

Like so very many human activities, progress in engineering design has developed through a very long period of evolution – rather than from revolution. Step changes do occur, but a revolution in thinking where the whole concept of thought has to be re-examined is rare. Today, however, we are already into a period of revolution (as the authors see it) but it is useful first to understand how the evolutionary period has led to today's designs. From this we can understand how the new multi-discipline approach to design can succeed, and we should be able to see the strengths and recognize the weaknesses of what has gone before and how it can help today's design technologies.

The industrial revolution was a step change, where the historical norms were changed fundamentally and in a relatively short space of time. Like all such changes, wealth was both created and destroyed, and a new order of life became the standard.

Historically, perhaps the present changes will be seen as the Information Technology (IT) Revolution. Certainly today the key to success is the rapid acquisition and use of information, and advances in all forms of IT are being exploited to acquire information. A good example was 'Black Monday', where financial manipulators were able to outwit whole national governments by being able to respond swiftly to changes in market conditions. The ponderous systems of the past could not respond quickly enough, and countries (such as the UK) lost fortunes. So it is with engineering. Today we can see the passing of the historical pattern of the engineering profession, with largely separate mechanical, electrical, electronic, and other groupings, to one where system engineering is needed to respond to the demands of the IT revolution in the outside world. Design is the key in engineering, and design has to react first to step changes in technology and demand from the outside world.

In looking back at the evolution in engineering design, we are not setting out to give an in-depth historical record, but to look at the basic tenets of design at different points in time, against the knowledge, materials, and indeed the economics of the day. Material sciences and metallurgy have always been very closely wrapped into what could be achieved at any one time, and examples of this facet are worth elaborating, to understand the interrelationships that always constrain our thinking.

In Chapter 1, we briefly mentioned that George Stephenson would not be totally surprised at stripping a modern multi-speed gearbox, as found in many cars, (Fig. 2.1). He might marvel at the compactness of the design, the precision of the components, and the manufacturing processes that could

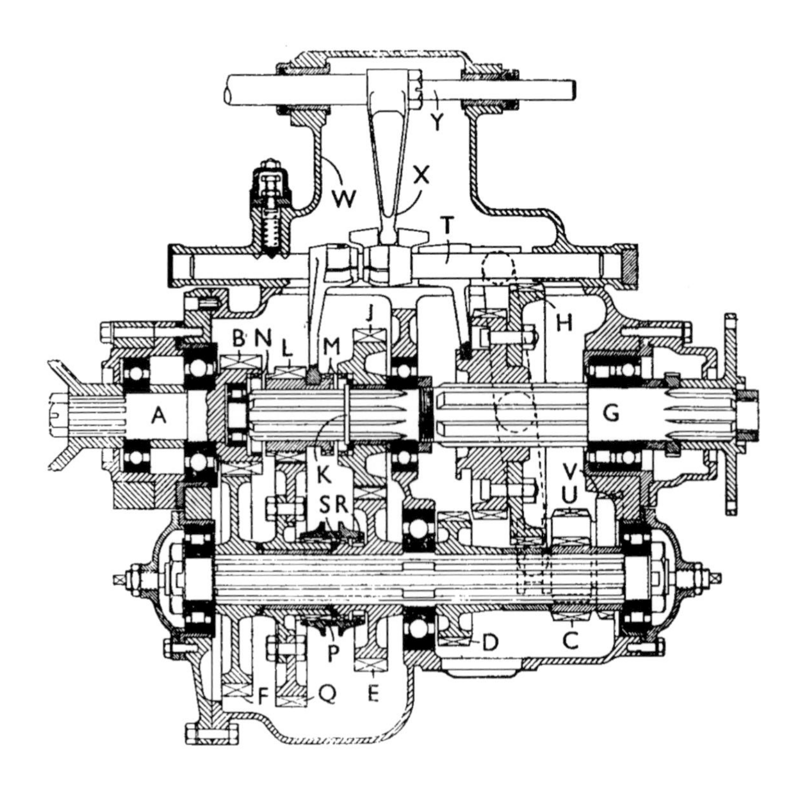

Fig. 2.1 Five-speed gearbox

produce such accuracies. He would, however, recognize three very salient design features:

- Gear ratios are dependent on the number of gear teeth on the gear wheels.
- Gear ratios are changed by sliding gear clusters along rotating shafts.
- Individual gear ratios can only be altered by physically changing the gears.

What this shows is that the basic principles of a gearbox were understood very many years ago; evolution has improved the manufacture and details, but the inherent limitations of the use of rigid gears remain, namely that to change the ratio you have to physically change the gear sets. Today, we can see computer-controlled electric drives where the 'gear' ratios can be changed at will, and in use, through software. This is what we would describe as the type of design 'revolution' that is currently upon us.

So what does evolution teach us?

2.2 Early design evolution

All early designs were what we might call empirical designs, mainly carried through piecemeal by the craftsman machine-builders themselves. It is very important to realize that these early machines were created by assembling existing types of sub-systems (in a systems engineering sense) into different configurations, to produce new products. An example may make this clearer. The first horseless carriages (an apt name) came from the addition of a stationary engine (a sub-system) to an existing type of carriage (another sub-system). The overall design had very simple mechanical interfaces between the sub-systems, but true integration in the control sense only came through the human driver. There was little or no influence of one sub-system on the other. It is possible to cite many other examples, such as the first steam-driven boats and ships, and the earliest steam locomotives. In all cases the new design was achieved with the addition of one simple sub-system to another or to others, with simple mechanical interfaces. It is the development of *interdependence* of one sub-system on another that is at the heart of the new design technology, and we will see that these interdependencies stretch further than just the engineering interfaces, whether these are mechanical, electrical, electronic, or control.

As early machines were seen to be successful, it was realized that the design principle of coupling various sub-systems together could be scaled up or down to achieve machines of varying sizes for different purposes. The earliest steam engines were examples of this type of integration and were used predominantly for pumping water out of coal and tin mines in England. This led to the publication of the first-ever design tables by Henry Beighton in 1721, which related the pumping capability of various steam engine piston diameters and the pump diameters needed to lift a given quantity of water from a given depth. These tables 'introduced precision into design calculations, which at that time were little better than guesswork' (**1**), and are perhaps the true birthplace of modern precision engineering design. As we look through the period from these early days to today, many actions stand out in a design sense, for instance:

- the early standardization of screw threads by Whitworth;
- the realization that standardization of sizes is important, first realized in the field of ordnance;
- the development of interchangeable parts and thus mass production in the early twentieth century.

In fact, the question is more what should be left out rather than what should be mentioned. There is one fact that demonstrates the importance of these early developments. As communications improved, each new technical development became the starting point for the next step. Communications here were not just by word of mouth or by the written word, but also by export of finished goods, so that new technologies could be viewed far and wide. Much of the development of the early steam locomotives was based on exported versions from England, which then developed locally to suit particular local conditions. What is apparent is that as each new technology came into being, such as steam engines, steam locomotives, motor vehicles, aircraft, military tanks, the jet engine, etc., then each new technology naturally started from the design principles and standards developed by the preceding technology (Fig. 2.2). This procedure also ensured that the simple assembling of the necessary sub-systems continued unabated until the later stages of the twentieth century.

It is easy to see this trend in the development of the motor car. Only some 25 years ago, cars were on sale which illustrate this principle admirably. The co-ordination and control across their 'sub-system boundaries' (brakes, steering, engine power, heating/ventilation) was solely through the human driver. Today we can see integrated systems between engine management systems, steering and braking systems, ventilation and heating/air-conditioning systems, and so on. Perhaps one of the earliest examples of some further sub-system integration was in the German Ford Taunus car, which in the 1960s combined the interior heating system with the engine-cooling system in a novel way. This was done so that the heating system was

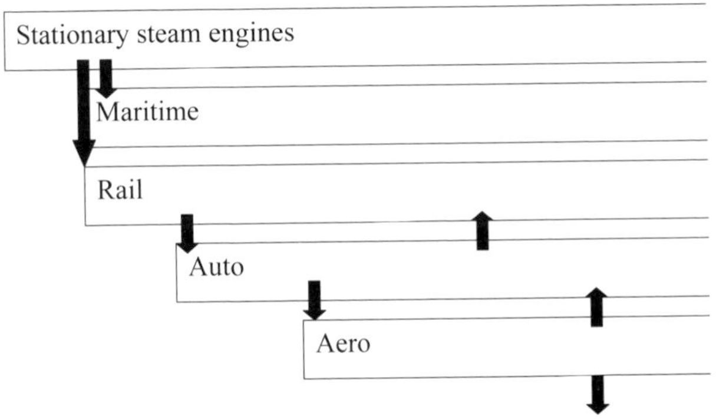

Fig. 2.2 Illustrative technology timeline

the primary system and normally took precedence, the engine-cooling requirements being automatically achieved by the ducting of cooling air through the interior heating system with a spill through the engine-cooling radiators, if that was required.

Similarly, many instances of modern integrated sub-system design came through military aerospace, where the control requirements of modern fighter aircraft passed the point at which a human could control the many sub-system demands at a sufficient rate, timewise. At the same time, we must not assume that automatic control of machines is a new feature. Speed control dates back to the earliest steam engines, through various types of governors. Indeed, the whirling balls of the 'Pickering' governor (Fig. 2.3) are an intriguing feature of the early stationary steam engine, but all these devices were designed to maintain a constant speed, something that humans are not good at controlling, particularly over anything but very short periods.

These devices were, of necessity, mechanical devices, and mechanical mechanisms are not the best means of achieving control because of the

Fig. 2.3 'Pickering' govenor

inherent inaccuracies in their manufacture. In fact, many of the limitations of mechanical control (or governing) have been overcome by the use of constant-speed electric drive motors, which in turn has led to an overt design philosophy of constant-speed machines. It is an interesting fact that many product-processing machines are designed to run at constant speed, even though it is known that product quality is speed dependent. Since the real output requirement from these machines is product at constant quality, rather than varying product quality at constant speed, it would seem obvious to design the machine to run at varying speed so as to achieve consistent quality. However, the design dogma of constant (and ever-higher) speed, originating from the days of mechanical governing, still prevails and thus this opportunity to design a higher-output machine is lost.

The use of constant-speed electric drive motors has, in fact, reinforced this constant-speed design philosophy across many engineering fields. At the same time, when we move away from electric drives as prime movers, such as in road, rail, and naval propulsion, the picture is slightly different. Where constant output is the norm, and thus constant speed of the prime mover is a requirement, the more efficient gas turbine can be used, as in aircraft, naval vessels, and electricity generation. Where variable speed is a requirement (as in road vehicles) the reciprocating engine holds sway.

In essence, what we are saying is that the co-ordination of unrelated simple sub-systems has been achieved until recent times through the actions of the human controller, whether driver, pilot, machine minder, or machine operator. The possibilities now open to us in engineering design, through true integration of multi-disciplinary technologies, will thus be predominantly through intelligent control systems, rather than purely through a human interface. This is a radical change in design philosophy. Let us study in depth a particular example. We will take as our case study the development of the horseless carriage into the integrated-systems motor vehicle of today.

2.3 Case study: Simple sub-system integration

If we start with a conventional horse-drawn carriage, of whatever age or time, we can see that the control over its speed and direction by the driver is a good example of intelligent control. The carriage is pulled by the horses (we will assume there is more than one horse, although it is not an important point), but the driver in effect gives signals to the horses as to their speed and direction. The horses are trained to accept and interpret these signals, and we can deduce that the horses must contribute something to the intelligent control process. The only physical effort required from the driver is in the braking mode, after he has signalled the horses to slow down. Of course, this system has a certain amount of uncertainty in the communication between the driver and the horses, and is vitally dependent on the training of the horses. If the horses have been poorly trained, or are high spirited (i.e. have independent minds), the driver may well have to exert added physical effort to control them and thus the complete horse/carriage 'assembly'. If we analyse this 'transportation module', we see we have a pulling or drive sub-system (the horses), a steering system (the driver with the horses, and pivoted front-axle assembly), a carrying module for the passengers, and a simple braking sub-system. Control is through the intelligence of the driver with some (indefinable) input from the horses. Only limited physical effort is required from the driver (in theory at least!).

The first mechanical engine capable of being made mobile was the steam engine, of course. Numerous early efforts were made to make viable steam-powered carriages, many with little success. The first was a three-wheeled affair made by the Frenchman Nicholas Cugnot in 1769 (Fig. 2.4). We might surmise that the new steering sub-system was not thought out too successfully, as Cugnot's vehicle hit a wall and he was arrested and imprisoned for being a danger on the highway!

Fig. 2.4 Cugnot's three-wheeled vehicle

The use of the steam engine at that period was probably most timely, since one of the great attributes of this type of engine is that it develops maximum torque at zero speed. This obviates the need for a clutch (which had not been developed at that time) and a gearbox, and provided a mechanical propulsion system akin to the characteristics of the horse, from a systems point of view. The disadvantages of the steam engine, using as it does an externally fired combustion system, actually outweighed the advantage of eliminating the horses, which is why the horse-drawn carriage survived until overtaken by the internal combustion engine at a much later date. Because of the disadvantages of the steam engine for true mobility (as distinct from programmed travel as on the railways), there was little real progress made in 'horseless carriage' systems development until the appearance of the internal combustion engine in the latter part of the nineteenth century.

At that time all the constituent sub-systems were available, even if in rudimentary terms. In fact, the appearance of the IC engine came when much progress had been made in engineering design and manufacture, and like the gas turbine at a much later date, it would probably have been impossible to develop the IC engine any earlier, due to deficiencies in materials and manufacturing techniques. From a systems viewpoint, the IC engine simplified the propulsion system tremendously, was much easier to control than the steam engine, and was less dangerous too – an important point in personal transport systems. However, it needed a clutch and a gearbox to transfer the engine's torque to the road wheels. These were new sub-systems, but they could be developed from existing technologies and added to the whole transport system (now called a car or automobile) in an add-on fashion. Probably for the first time, the human driver had to exercise not only intelligent control but also real physical dexterity in using the clutch and the brakes, with a degree of co-ordination not required previously. Some early designs simplified the process a little, such as the use by Ford of a two-speed epicyclic gearbox in the famous Model T (Fig. 2.5). For most, the horseless carriage embodied this early state of development; note that it coincided with the first effective need for the human controller (or driver) to exercise new manual skills.

Fig. 2.5 Model T-Ford

We can jump forward now to the post-WW2 car, which is really a much-refined version of these early examples. We see the various sub-systems in developed form, but still with simple mechanical interfaces. If we now look at the introduction of computer science to the car in the last 20 years of the twentieth century, we see firstly that improvements were made in the individual sub-systems (electronic ignition, engine management systems, and so on). New microcomputer based systems were then introduced (anti-lock brakes, traction control, etc.), and sub-systems integration moved to electronic interfaces. This process is still in progress, and more and more systems are being introduced to make the driver's task easier (or should we say to de-skill the driver's task?). If we look at the extreme example – Grand Prix cars – we see traction control, launch control, clutchless gear changing, and active suspension; it is readily accepted in Formula1 that races could be run without drivers, like a full-size Scalextric race.

If we step back and analyse what has happened, we might say that the early development of the car was based on the integration of separate mechanical sub-systems, with very simple mechanical interfaces, necessitating increasing skill on the driver's part. Later development has had two effects. One is that engine efficiency has been improved through programmed engine control, and the other is that the driver's task is getting easier. Is the added complication worth the improvements? Do drivers want their task to be de-skilled? Is it *safe* to de-skill the driver's task in this way? What are the full-life implications of these multi-disciplinary designs? Overall, what is the balance between the improvements and cost to buy, cost to run, and total reliability? These considerations must be the main *starting points* for our design thinking today. In particular, is our knowledge sufficiently comprehensive to allow us to program intelligent control systems to cover all likely contingencies? Do we need to move towards self-learning control philosophies, rather than relying on our own information and experience to pre-program standardized controls? While the new technologies are giving us new answers and new approaches to design, all the parameters are not yet clear, so we must proceed with caution, or at least with knowledge of the full implications. An example will illustrate this point.

2.4 Case study: Vehicle ignition systems

As we have seen, the earliest cars were little more than a stationary IC engine applied to a cart or carriage. Engines for stationary duties are usually governed and are certainly not best suited for variable speed running. While the throttle is the main control on engine speed, ignition timing is also an important factor. Retardation of the spark for easy starting and advance for high-speed running were recognized from early days, but the means to achieve these were limited. Early cars had a hand ignition-control lever on the steering wheel and it was left to the skill of the driver to judge the correct amount of advance. At least the 'closed loop' of this means of control is fairly obvious and close at hand! It is unlikely that the engines ever ran at the correct ignition advance settings; we now realize that the required ignition advance is a function of several related factors, such as engine speed, power requirement, ambient temperature, and indirect factors such as throttle position.

The introduction of the mechanical contact breaker with its own advance and retard mechanism was seen as a significant development. The mechanism could be set mechanically to a given static advance figure, and various straight-line-type advance curves could be obtained by varying spring strengths, bob weight size, and end-stop position (Fig. 2.6). Some offset against throttle position, measured via intake depression, could be introduced through an evacuated capsule, but the mechanism was basically a control against engine speed.

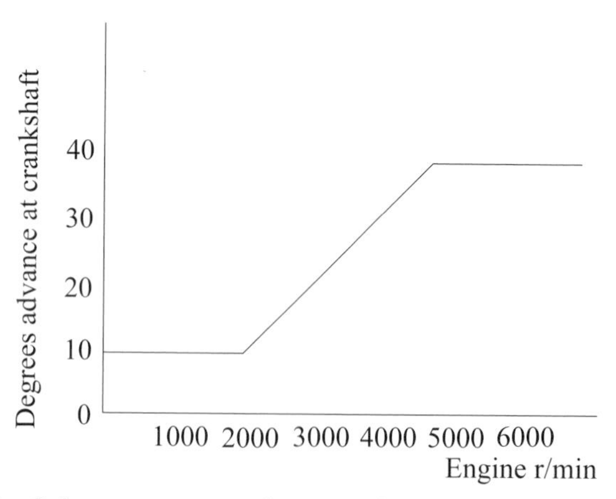

Fig. 2.6 Advance curve for mechanical ignition system

This type of system prevailed right up to the late 1970s, when electronic systems started to take over. The mechanical system had been developed into a very reliable unit, whose occasional need for maintenance (resetting of the contact-breaker gap) was chiefly caused by erosion of the contact points due to the switching of quite high electrical currents. Consequently, the first electronic systems still used the mechanical contact-breaker system but now merely to trigger electronic switching of the firing currents. This improved the reliability of the system still further, but did nothing to improve the advance curve characteristics of the unit.

Full electronic control of the ignition came later, and typically uses a 16 × 16-bit map driven from engine speed, throttle position, and ambient temperature sensors. In mapping these types of system, it has become apparent that the old straight-line ignition curve is a long way from the optimum, and that precise timing of ignition has only recently become possible, typified in Fig. 2.7. The system is also admirably suited to interfacing with full engine control, as seen in modern cars.

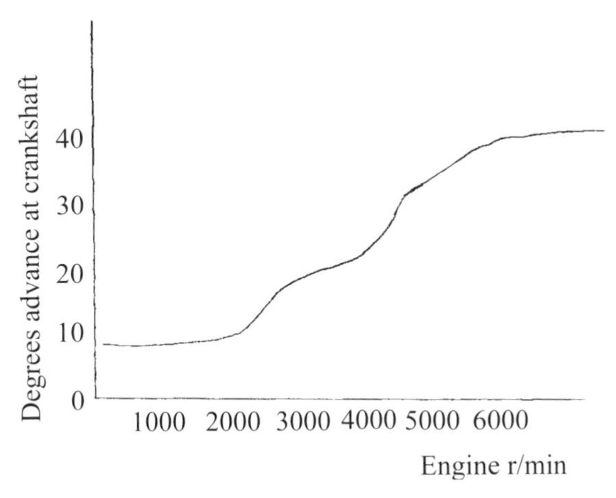

Fig. 2.7 Advance curve engine requirement at one given load

Another similar case study showing how existing technology has reached a plateau in design is outlined in Section 7.1 (The wrapping machine). Both these studies show how quite sophisticated sub-systems have developed progressively within the bounds of a mechanically based system, until an absolute limit has been reached with that particular design approach. In general we have seen how design started from simple experimental assembly of existing sub-systems, increased control through the human interface, and increased speeds through reduced inertias, until in most areas a limit has been reached which cannot be breached by conventional design. Costs, weights, and power requirements have all been increased to gain improvements. Heat rejection, recycling, and green aspects have not been a powerful consideration. Optimum performance (however measured) has not been achieved with simple mechanical sub-systems, although they have provided adequate performance and good reliability.

2.5 Modern technologies

The impact of the computer is now all around us and it is becoming difficult to cite cases where its influence is missing. In mechanical engineering the biggest impact most probably has been in computer control of mechanical devices and the use of computing in design itself. The term 'mechatronics' is used to describe the designs that incorporate electronics with mechanical engineering; it is a term that is recognized across the world. We know that the influence of mechatronics is erasing the traditional boundaries between mechanical and electrical engineering and this influence can also be seen strongly in the civil engineering/construction industry sector, medicine, marine engineering, and virtually any other branch of engineering you might care to define.

2.5.1 Control

We are passing through the last stage of a phase in engineering where the use of electronic controls has been little more than an 'add-on' to existing machines. This book is about the true integration possible with mechatronics, where the fundamental design philosophies are vigorously re-examined. The addition of an electronic control system to an existing machine design is not true mechatronics, since the added potential of the integrated design will always be greater than adopting this more simplistic approach.

This is not to say that the simple add-on system does not have its place, and in many cases the electronic control will confer benefits, as has already been shown in the case study on ignition systems. However, an element of control is a definitive requirement in any mechatronic system and can be defined as the integrating factor for what is mechatronic when allied to input sensors and transducers. At the lowest level of complexity there might be cyclic operations dependent on inputs from sensors with output to controlled actuators. The programmable logic controller (PLC) is adequate for this type of simple sequential control, and the PLC field has improved tremendously over recent years after a rather confused start. The use of transducers for measurement of position, force, temperature, speed, etc. really necessitates the use of feedback loops if the full potential of the system is to be realized. Control is then frequently through the use of some form of a proportional, integral, derivative (PID) controller, using techniques that can give a digitally implemented algorithm which can then be used by a microprocessor-based controller.

An important added facility with many of these control systems is the ability to provide and display process information and diagnostics. While statistical process control (SPC) has been in use for many years, the provision of extra sensors and transducers together with time-measuring facility has updated this capability and also enabled the machine to become self-correcting. The Molins OASIS system was one of the first to exploit this capability, timing the intervals between the instances when different sensors and transducers fired off to derive a stoppage or malfunction message for the operator, and thus providing added management information into the management information system (MIS).

Current control systems have deviated from numerically based, data-dependent control algorithms to those using knowledge-based rules, which now include fuzzy logic and expert systems. These knowledge-based systems are aspects of artificial intelligence (AI), which is the approach needed when the machine operates in a field of uncertainty. Intelligence may be defined as the capability of a system to achieve a goal or sustain desired behaviour under conditions of uncertainty. Information is a means of reducing the uncertainty about a domain, so by these definitions machines need information (via sensors and transducers) to have sufficient intelligence to operate under uncertain conditions. Machines and systems (with their controllers) may thus be defined as falling into a number of categories:

- Programmed (predictable; operate in predictable fields)
- Proto-intelligent (mostly predictable)
- Intelligent (unpredictable) – showing autonomous behaviour, self-diagnosing and self-repairing, self-learning and self-organizing, with ability to negotiate with other systems and possibly eventually be creative.

The next stage towards intelligent behaviour appears to be multi-agent technology, or multi-agent systems (MAS) as they are known, which have the capability of dealing with many more variables and producing self-learning and self-diagnostic/self-correcting systems. The author and others are currently exploring these types of systems in research into the Intelligent Geometry Compressor, where upward of 500 variables will be controlled in an autonomous system. This type of approach is covered in more detail in Chapter 5.

2.5.2 Sensors and actuators

Sensors and actuators are the key to modern mechatronic systems. Because of this there has been much research carried out in this sphere and this is continuing today. However complex the system and however complex the control approach used, the system will depend on sensors to tell it what is happening and actuators to make any required final movements. This is not the place to give an in-depth treatise on modern sensors or actuators, but merely to point to a number of approaches in both the devices themselves and the different ways to employ them. A more in-depth look at how to make engineering measurements (which is just what sensors are there to do) is contained in the book *Engineering Measurements* (**2**).

Many sensors are used to measure parameters on a continuous basis, while others are used to 'fire off' when a particular event occurs. An obvious example of the latter is a fire alarm. The sensor is required to remain dormant until a certain type of situation arises, in this case possibly a sudden rise in temperature, when the sensor then actuates an alarm bell. With this type of sensor there is an important aspect that can be overlooked – the information that can be deduced by looking at both the sequence and the time intervals when a batch of sensors fire off. In a processing machine, for instance, the product flow will be monitored by different types of sensors at different parts in the flow path. If the process breaks down, the operator can be warned of the likely cause of the problem by analysing the pattern of sensor activation *and* the time intervals between each activation. This type of approach is essential for self-learning systems and has great potential for future intelligent control, as well as fitting in well with management information systems (MIS) and statistical process control (SPC) requirements.

An example of the former type of application is in the axial compressors used in the majority of gas turbines. This type of compressor will contain many rotor blades and stator (static) vanes. With sufficient sensors on the vanes (and possibly in time on the blades as well), local aerodynamic instabilities can be detected on a continuous on-line basis, enabling continuous vane geometry corrections to be made to reduce the instability. This is the basis of the Intelligent Geometry Compressor, now under study.

The developments made in controlled motor drives are also mainly applicable to actuators, particularly through applications using linear motors. Actuators are probably the most important area for development in mechanical engineering at the present time.

2.5.3 Power supplies

A feature that is becoming apparent in recent systems designs is the need for higher-powered electric supplies, which necessitates a reappraisal of the conventional norms in applications such as the military, motor vehicles, and aerospace. While in self-contained systems the use of typically higher-voltage electrical systems (moving say from 12-volt systems on cars to 24- or 42-volt systems) undoubtedly tackles the local problem, on a systems engineering front it raises serious questions as to compatibility, maintainability, and possibly disposability. The increase in use of complex control systems, with their need for more sensors and more complex actuators is an inevitable advance which necessitates increased power requirements; however, the introduction of a step change of this type needs very careful control, perhaps even at an international level. There are so many case studies that can be quoted about step changes in technology being incorporated with insufficient forethought, that a change of this type needs very careful control.

2.5.4 Controlled drives and mechatronics

The development of controlled drive motors and actuators has produced a situation where displacement and velocity profiles can be devised to best suit the task in hand and then be specified via the appropriate software. Displacement and velocity profiles obtained via mechanical means, whether with cams, gears, linkages, or Geneva mechanisms and their like, always introduce compromises in requirements, and in the end also necessitate time-consuming and expensive remedial work to change these profiles in any way. One important attribute of the controlled drive approach is therefore in speed of change, as well as the improved accuracy in obtaining exact controlled movements.

Different types of motor, actuator, and software are best suited for different tasks, but it is a complete move away from the constant-speed drive motor and complex mechanical systems used to obtain approximations to these required profiles. Much work remains to be done in this field – which, while maturing, is still in its infancy – if our predictions of the future are to be correct. However, it has to be accepted that it is very difficult to imagine a return to the old 'mechanical' type of solutions, mentioned above. This field has been researched very vigorously for over 15 years, and continues to be, particularly in the UK through the Design of High Speed Machinery (LINK) research programme (**3**), which brought together industry and academia in a unique effort to develop the technology into a workable technique. A useful small information booklet where the characteristics of these types of motors are discussed is *Electrical Motor Drives for Mechatronic Systems*, published by the IEE (**4**). This gives a useful introduction to the field, discussing the merits and characteristics of different types of drives and controllers.

The use of stepper motors is fairly well known; they are a popular drive force for low-power position control with high holding torque. However, brushed permanent-magnet and wound-field-type motors are also being widely used. The use of the more modern, brushless AC and DC motors gives a controlled electronic drive system that avoids the limitations of physical commutation in the motor. These are ideally suited to servo applications, because the linearity between current and torque is maintained, as in the brushed type of motor. Brushless and induction types of motor are suitable for both rotary and linear applications, while brushed DC motors, reluctance motors, and stepper motors are limited to rotary applications. Brushed DC motors are popular in consumer products.

This whole technology is still developing very quickly, particularly as to the torque and power characteristics that are obtainable from these motors. There are, however, both power and size limitations at present and there is continual development to push these limitations forward, so reference to the commercial makers is recommended to find the current position. On the control side, closed-loop control is needed as a minimum for a robust system, and it is in the control area that the largest advances are likely to come. Control and embedded intelligence is discussed fully in Chapter 5. Figures 2.8 and 2.9 show pictorially how this approach can be applied to a specific case, in this example to the wrapping machine studied in Chapter 7 (Section 7.1). In a more diagramatic presentation, Figs 2.10 and 2.11 show how the control function passes from fixed geometry mechanical interactions to controllable software functions.

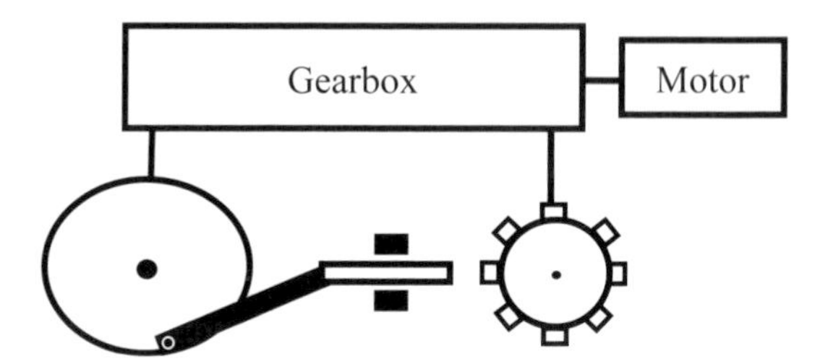

Fig. 2.8 Phase synchronized drives

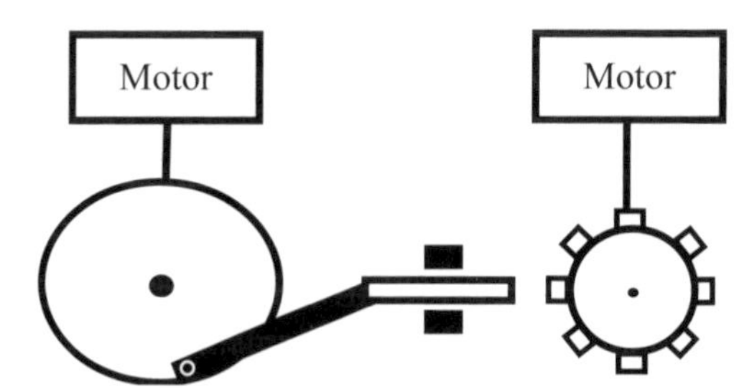

Fig. 2.9 Phase synchronized drives – new approach

It has to be remembered, however, that as we develop existing or new drive technologies, the final actuation must be through a mechanical interface, as we operate in a physical world. We can always debate whether magnetic levitation is a 'mechanical' type of actuation or not, but the principle is clear! Mechatronics is the science that covers these cross-discipline technologies, and to date there has been a successful integration

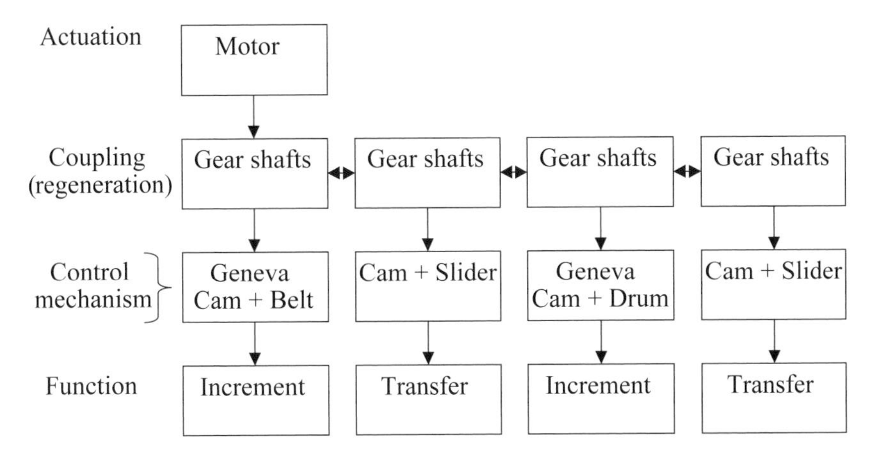

Fig. 2.10 Traditional approach to machine design

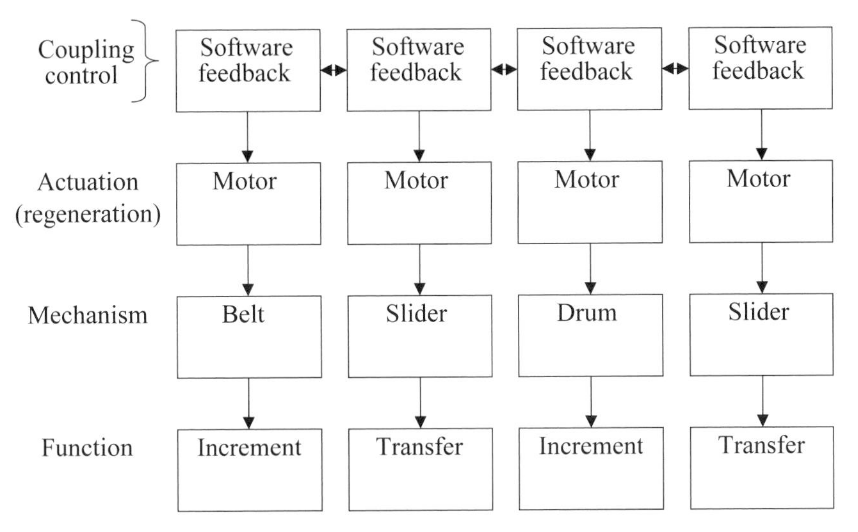

Fig. 2.11 Independent actuators approach to machine design

of different fields, mainly mechanical and electronic disciplines. As in many walks of life, it all depends on the definition, and the Western world has had more difficulty with the exact definition of mechatronics than has the Eastern world, where the term originated. The definition seems abundantly apparent in the name itself, however.

We have shown how engineering design historically has followed an evolutionary path, with the result that many tenets of design are no longer questioned. Perhaps another simple case study here will illustrate this point more clearly, and provide a vehicle to look at how we might redesign a simple mechanism with the new technologies.

2.6 Case study: Force versus timing

Reciprocating engines are a prime source of power generation and have been in use in one form or another for approximately 200 years. As distinct from turbines, all reciprocating engines need some type of timing device to control the transfer of fluids into and out of the working chamber. In the most common forms of these engines, the fluids are gases and the timing device is a mechanical type of system (petrol and diesel engines, steam engines, and so on).

The objective of any valve gear or valve driving mechanism must be to control the timing of the inflow and outflow of the working fluid through the cylinder, to suit the particular load and speed requirements of the engine. Hence, the valve timing required at any particular speed and load can be specified and in most cases might require instantaneous maximum valve opening and instantaneous valve closing at the required points in the working cycle. In a diagram showing valve opening versus crankshaft turning, it might take the form of a square wave (Fig. 2.12). While this theoretical displacement (valve lift against time) diagram may be modified to suit other considerations (such as noise, for instance), basically the requirement is a square wave.

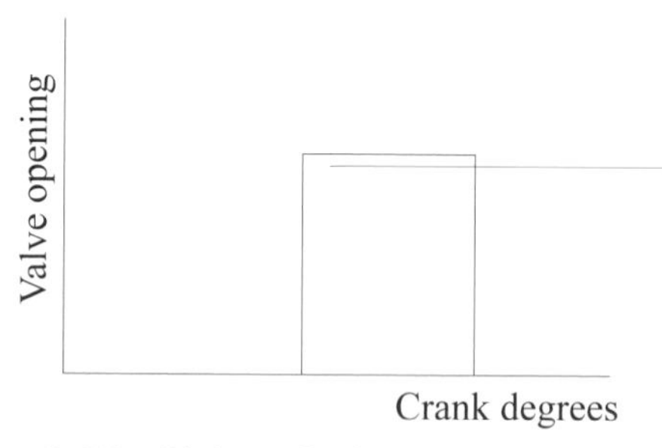

Fig. 2.12 Valve timing-square wave

The ingenuity that has gone into the design of practical valve-timing devices is quite amazing, particularly where there is a need to vary the timing of valves while the engine is running and under load. The most common type of valve gear design today is that used in internal combustion engines, which are generally of the cam-operated fixed timing type (Fig. 2.13). A good example of a variable-timing type of valve gear is that used on a steam engine, particularly the old steam locomotives (Fig. 2.14).

If we examine any of these valve-driving mechanisms, we find that there is a common characteristic across all of them. A mechanical system is used which is required to carry out two functions simultaneously: to provide the driving force to move the valve, and to provide the precise timing of the

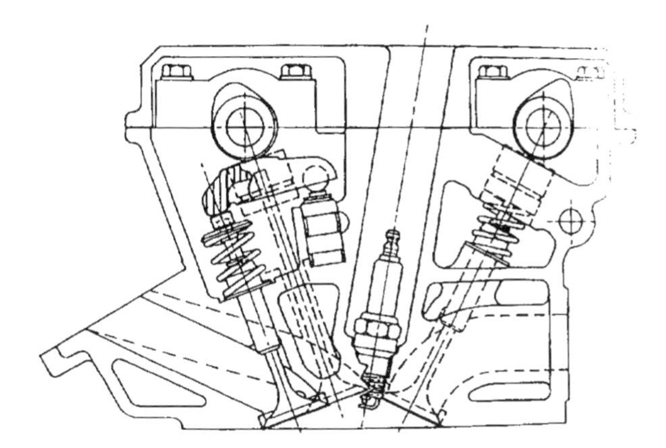

Fig. 2.13 IC engine cam/valve drive

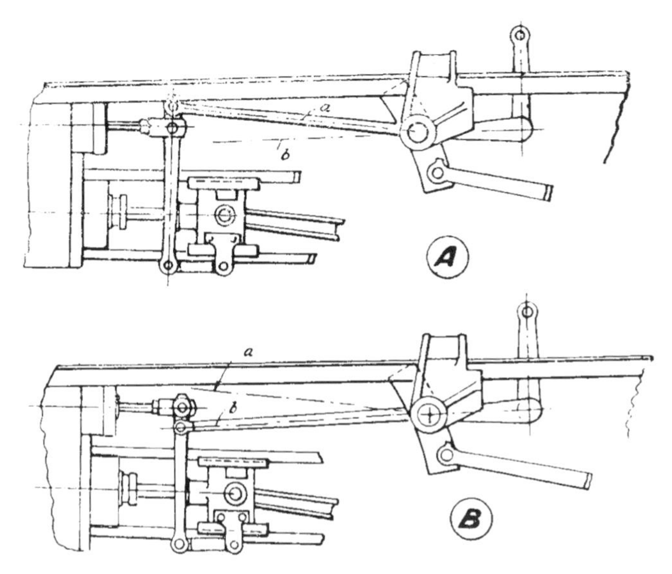

Fig. 2.14 Walschaerts' valve gear

valve to supply the required transfer functions. The important point to note is that this dual requirement compromises the individual functions required; for instance, in an internal combustion engine the timing of the inlet and exhaust valves has to be compromised by the accelerations and decelerations mechanically possible with a cam-operated system. Similarly, the valve timing on a steam engine is related to the geometric movements of the mechanical linkages used, rather than to the precise timing of the steam inlet and exhaust.

It is instructive to analyse the valve-actuating system from a systems engineering viewpoint. First we would need to define the exact timing of both the inlet and exhaust functions to optimize performance, disregarding the constraints imposed by the operating system. We would do this for a range of different engine speeds and loads; a commonly accepted range would be a 16 x 16 map. We would then define the driving force required by the valve, and its variation (if any) throughout this range of the operating cycle. Can the timing requirements be achieved with just one valve, or is a multi-valve approach needed? We would then see if any single operating system could achieve all the requirements (timing and drive) simultaneously, and quickly conclude that they most probably could not. At this point there is a choice between the single system that compromises one or both functions (as in current operating systems), and a system that uses dedicated sub-systems that each provide the correct individual requirements.

A systems engineering approach would be to separate the valve-timing function from the valve-driving function, using say computer-controlled linear motors or a hybrid mechanical/electrical or hydraulic actuator. In all systems engineering applications the importance of the final actuator is paramount. With the advent of computers, it is interesting to note that numerous programs have been devised to optimize the compromised systems presently used, but little if any attempt has been made to adopt a systems engineering approach and to separate the valve functions. Once a choice of sub-system has been made, we then need to make a vigorous comparison with alternative systems and with compromised systems (such as those existing today) before making a final choice. In this final choice all other factors have to be considered, such as cost to make and cost to maintain (whole life costs), reliability, training necessary for changed systems, suitability for different environments, and so on. The key, however, is the starting point, where the fundamental processes to be controlled are reassessed against the existing technical solution and compared with a modern mechatronic approach.

2.7 References

(1) **Briggs, A.** (1982) *The Power of Steam*, Chapter 4, Bison Books, University of Nebraska Press.

(2) **Polak, T. A.** and **Pande, C.** (1999) *Engineering Measurements – Methods and Intrinsic Errors*, Professional Engineering Publishing, Bury St Edmunds, UK.

(3) **Department of Trade and Industry** (1997) *The Design of High Speed Machinery (LINK) Programme*, March, DTI/Pub 2671/6k/3/97/NP (Crown Copyright).

(4) **PGI 6, Manufacturing Division, Institution of Electrical Engineers** *Electrical Motor Drives for Mechatronic Systems: Information Sheet.*

Chapter 3

The Need for a Systems Engineering Approach to Engineering Design

3.1 Introduction

Engineering design is fundamental to the engineer's contribution to society – providing solutions to real-world problems. Increasingly these problems reflect the complexity of the modern world, characterized by increasing levels of integration, a wider range of criteria to be satisfied, and accelerating change. In short, we live, work – and engineer – in an increasingly 'systems' world, and a systems engineering approach is essential if we are to meet this challenge.

In this chapter, we discuss the nature of the engineering design process and the systems engineering approach, as traditionally expressed, and expose the deficiencies in relying only on a systematic approach.

3.2 A traditional view of engineering design

The 'engineering design' approach has always recognized the central importance of knowledge and expertise in the fundamental issues with which the engineer is dealing, whether it be mechanical structures, electronics or building materials, etc. Over the years, the sophistication and range of technologies associated with individual engineering domains and the associated body of knowledge have, of course, increased dramatically. The professional engineering institutions are central to ensuring that such domain expertise is communicated, taught, understood, and implemented in a consistent, professional manner. In addition to the engineering domain knowledge, there has been recognition for many years that there is also a 'process' aspect to engineering that transcends the separate domains of technological expertise – that 'the engineering design process' itself has fundamental, generic characteristics. This recognizes that engineering, of whatever domain, is about creating and producing designs that provide successful solutions to real-world problems and that there are some fundamental principles that are applicable to any technological and application area of engineering. These fundamental principles are invariably expressed to encompass aspects such as (**1**):

- Clearly establish the primary need; that is, express the real-world problem in clear terms.
- Analyse the problem and generate several potential solution options.
- Evaluate the different solution options and compare them to establish their relative advantages and disadvantages in satisfying the primary need.
- Select the design solution that provides the most acceptable solution to the primary need.

An important aspect of the approach is to ensure that one has a sound basis for engineering decision-making, from a clearly understood statement of the problem through to clear criteria on which the evaluation and selection of potential solutions will be made. This is commonly referred to as a 'systematic approach to design'. It is usually recognized in the engineering design texts that only in the simplest situations will such a process be conducted as a single, forward pass. In realistic, more complex situations there will be a need to 'iterate' around aspects of the process (or even around the whole process) as issues are explored and new dimensions to be considered are discovered. This is illustrated in Fig. 3.1. Hence, for example, design for manufacture, design to cost, etc., all involve setting out some initial view of potential solutions, exploring their implications and then iterating to establish modified views based on the understanding gained.

The design process iterations shown in Fig. 3.1 serve two purposes.

- First, they reflect the learning process, with original propositions being explored, their implications identified, and revised propositions being formed as a result of this new insight into how different characteristics drive different aspects of the outcome. Design for manufacture is an example of this, where an initial design proposal is assessed in terms of its manufacturing implications and, where possible, modified accordingly.

- Second, the iterations reflect the need to revisit earlier (provisional) decisions as the need to trade off some aspects of the original intentions is discovered. So, for example, the original statement of the problem may have set out some performance, time, and cost requirements that may later have been found to be mutually incompatible. Trade-off decisions would then be necessary to decide whether, for example, to increase the cost or reduce the performance requirements.

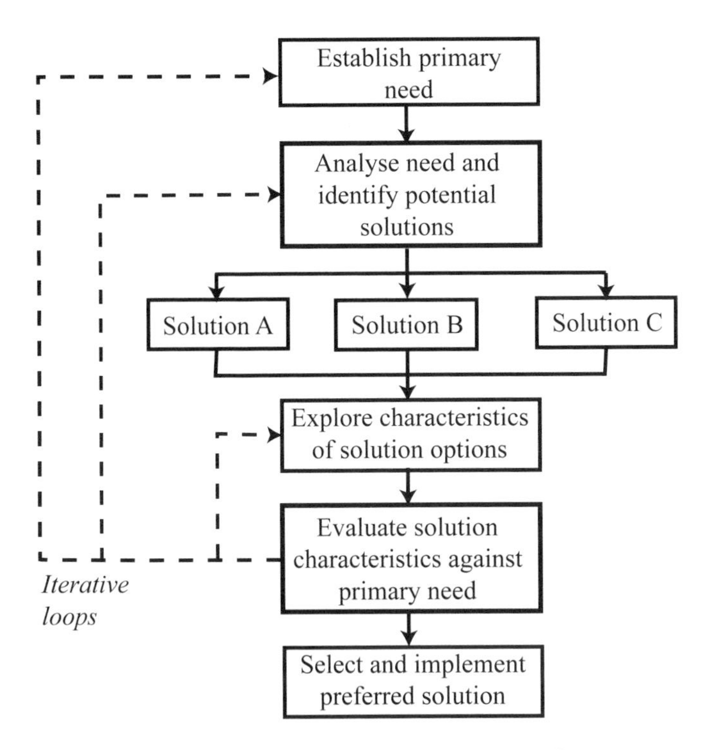

Fig. 3.1 A fundamental engineering design process

The expression of this generic engineering design process encompasses many features that are now commonly accepted best practice:

- A 'systematic' approach in which the problem is stated clearly, and several potential solutions proposed and evaluated against the problem, with a preferred solution being selected from the process.
- An iterative, exploratory approach, with techniques such as simulation and prototyping providing useful means of exposing the characteristics of the solutions being assessed.

- A multi-disciplinary team approach, with the viewpoints and expertise of a range of disciplines being involved in the decision process. The disciplines involved must encompass all the issues that are considered to be important considerations, including, for example, through-life 'stakeholders' such as users, production experts, maintainers, and so on.

3.3 A traditional view of a systems engineering approach

There are a vast number of textbooks that describe a traditional view of a systems engineering approach. This has developed over the last 50 years or so as a formal, disciplined approach to the engineering of systems, and it is widely recognized that we are dealing more and more with 'systems' and so must take a systems engineering approach. But what do people usually mean when they talk of a systems engineering approach?

As with the engineering design process discussed above, the systems engineering approach is recognized as needing to be applicable across any technology area and a wide range of application areas and so it must be expressed in generic terms. The International Council on Systems Engineering (INCOSE) has produced several documents that set out the principal features of such an approach. For example, INCOSE has published its 'Pragmatic Principles of Systems Engineering' (**2**) as follows:

(a) Know the customer and the consumer.
(b) Use effectiveness criteria based on needs to make decisions.
(c) Establish and manage requirements.
(d) Identify and assess alternatives so as to converge on a solution.
(e) Verify and validate requirements and solution performance.
(f) Maintain the integrity of the system.
(g) Use an articulated and documented process.
(h) Manage against a plan.

These principles relate very well to those discussed above for the engineering design process, as illustrated in Fig. 3.2. The INCOSE Pragmatic Principles emphasize very much the same overall process and also the importance of good engineering/managerial practices, such as documenting and communicating the whole process, its decisions, and its products. We can see that the main emphasis of this widely accepted view of a systems engineering approach is to ensure that the generic engineering design process is employed successfully within systems programmes.

Over the years, many of the activities associated with defining and

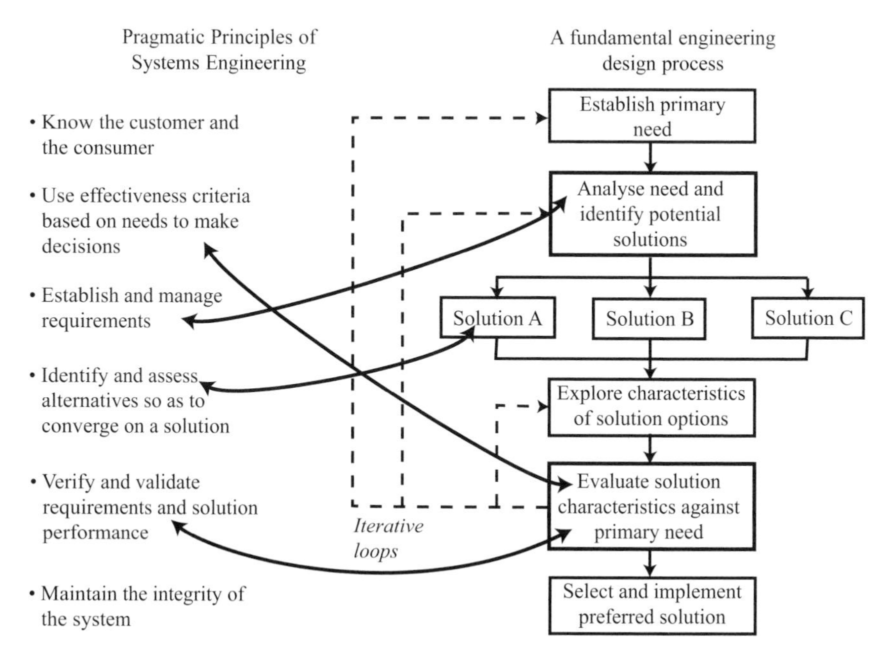

Fig. 3.2 Fundamentals of a systematic approach

improving the systems engineering approach have been focused on developing (sub-)processes, methods, and tools for making this process work effectively and for managing it. There are many textbooks in systems engineering (for example, **3**) that expand on the context of this fundamental process and amplify many of the issues that need to be tackled successfully within such an approach, such as:

- the overall development life cycle and the conduct of the fundamental process within it;
- the role and conduct of integrated product development teams;
- the importance of taking a whole-life approach throughout the process;
- systems engineering management plans;
- design reviews and audits;
- configuration management;
- requirements capture;
- functional analysis;
- 'Speciality engineering', such as reliability, safety, etc.;
- and many other such aspects.

Many texts (for example, **4**), describe approaches and notations for particular aspects of the process, such as:

- categorizing and expressing user requirements;
- categorizing and expressing system requirements;
- expressing system architectural design.

All these texts deal with very important issues and provide useful discussions on sub-processes such as systems engineering management and requirements engineering. They represent the 'traditional view of systems engineering' and there is a great deal of activity underway associated with developing and defining approaches to the issues they tackle. For example, a popular issue at the moment is the question of how best to use an approach such as the unified modelling language (UML) within the process. This particular example demonstrates a further point: that many of the processes, methods, and tools that are applied within this systems engineering approach originated in software engineering and have then been applied to 'systems' in a more general sense. This is not altogether surprising or inappropriate since systems are, like software, by definition more *abstract* than hardware components and so cannot be expressed using well-understood notations such as those associated with the laws of physics.

We can see a consistent, systematic approach underpinning all these process views. This consistent systematic approach is based on many fundamentally sound aspects such as: ensuring that one is clear about the problem one is trying to solve; ensuring that one does not presume a solution at the outset; involving many disciplines in the decision process, reflecting the need to expose problems and correct them sooner rather than later. However, while such a systematic approach is widely regarded as correct, there are several criticisms that are often aimed at it, including the following:

- 'It's only common sense.'
- 'It's what good engineers have been doing for years.'
- 'It's only management.'

These criticisms are difficult to refute entirely. The fundamental process *does* look like common sense – after all, who would commit to a solution *before* the problem has been identified! Unfortunately, as we know, common sense is not always that common and there are many instances of this very 'crime' being committed. Similarly, it is probably true that 'good' engineers have been doing this for years, but that does not mean that there is no merit in trying to capture best practice and encourage *all* those involved to employ the approach. It can be of great benefit even just achieving a common understanding of the objectives and decisions across the multi-disciplinary team, since misunderstandings can, of course, undermine the whole process. So, while there is an element of truth to these common statements about the process, it is wrong for them to be made in a critical way, as if the process should not be employed.

However, even though these common 'statements' do not in themselves invalidate the sort of systematic process described above, there is another major limitation in this traditional view of the systems engineering approach. It does not emphasize sufficiently the need to establish the systems one is dealing with, to ensure that the process is applied to a robust, meaningful understanding of the problem and its solutions. It is important that the decisions made throughout the process (such as the evaluation and selection of the solution) are made on the basis of robust assessments of the real-world issues. This 'systems understanding' may be obvious when we are dealing with limited, well-understood situations but that is rarely the case, particularly in the modern world, and so it is becoming more and more important to redress this deficiency in the traditional view of systems engineering. In the following section we discuss how the issues that impinge on the systems engineering decisions are becoming much more wide-ranging and diverse, and how the relationships between them are becoming much more complex, making it increasingly important to understand how to achieve confidence in the decisions being made.

3.4 Systems and systems influences in engineering design

The world is becoming much more 'joined up' and many things that we once dealt with in isolation now interact with each other more than ever. As we follow the fundamental principles of the process described earlier, it is very important to be able to understand the elements and issues we must deal with, the relationships between them, and the overall 'behaviour' of the total system. This is certainly true in engineering and the issues with which it must deal. Some examples follow.

- Technologies and engineering products are much more highly integrated than ever and invariably the characteristics delivered by modern engineering products are achieved not by a single technology but by a 'system' of several elements and technologies operating together. This in turn means, of course, that the engineering disciplines must work together within the design process in a similarly integrated manner.
- The relationships between the engineering products and the environments in which they are used are also becoming more complex, and these relationships take many forms – from interactions with other systems' external technologies and people, to influences of

legislation, and so on. This presents us with a highly complex picture to develop and understand.

- Even the organizations in which the engineering process is undertaken are more complex than ever. Increasingly, the integrated teams that work together in the process are distributed across different enterprises and across the globe. This makes it even more important to understand the interdependencies and the critical issues within the process so that the communications, information sharing, etc. across the organization can be designed to enable rather than hinder robust decision-making in systems programmes.
- The engineering domain processes, methods, and tools within the overall engineering design process are increasingly also highly integrated. Previously, such domains were effectively separate and operated in an 'over the wall' way within a sequential engineering design process. With the increasing emphasis on reducing time and cost and the effective operation of integrated, multi-disciplinary teams, the relationships between these processes, methods, and tools must be fully recognized, and employed as effectively and appropriately as possible. For example, enterprises are investing heavily in developments such as shared data environments, not least to provide a communications infrastructure across global operations; but such innovations can deliver their true potential only if they are used appropriately, with the critical information being in the right place at the right time, and with the associated decision process recognizing the interdependencies between issues and contributions from different domains.
- The engineering design process view emphasizes that the evaluation and selection of solutions must be made against their ability to provide an effective response to the real-world problem. Given that we are faced with the 'joined-up' influences mentioned above, and that the real-world customers are increasingly not content to merely take delivery of engineering products but instead want to be convinced that they offer an effective solution to their needs, the decisions must be based on an evaluation of the behaviour of the engineered product when it is used in the wider real-world environment. Once again, this demands that we develop a clear, robust understanding of the total system influences that must be taken into account.

We can see that these examples cover a wide range of different issues, but they all have the same fundamental issues in common: they are all examples of systemic issues and relationships that are critical to the success of the engineering efforts. Develop a flawed understanding of these 'systemic' issues and relationships, and the traditional engineering design process/systems engineering approach described earlier will not achieve a solution that will deliver an effective resolution to the right real-world problem. In short, we risk solving the wrong problem in the wrong way, and making poor, ill-informed and inappropriate decisions.

So, while the traditional and widely documented 'common sense' process described earlier expresses the fundamental principles of the process itself, we also need to find ways of understanding 'systems' and their behaviour in the real world, and we must be able to address, as engineers, questions such as:

- What is the problem that we are trying to solve? What is the overall objective to be achieved? What issues and influences does this present us with and how are they influenced by different solution options?
- How can I establish a view of the problem and the real-world environment that is correct and robust? How can I judge whether I have considered all the issues and correctly identified the critical aspects?
- How should I evaluate proposed solutions so as to base decisions and trade-off judgements on the real world effectiveness, not on the design features themselves?
- Who needs to be involved in the engineering process, when, and for what decision? How can I judge which decisions are critical?

Earlier, we mentioned that a common statement made in relation to the fundamental (traditional) systematic approach is that 'it is what good engineers have been doing for years'. However, it is important to realize that the complexities associated with the issues mentioned above did not have to be tackled explicitly by 'good engineers' in the past. The situations, problems, issues, and interdependencies that they dealt with were well understood and they had developed a great deal of experience in dealing with them. Now, we can no longer take these issues for granted and we must establish the 'system understanding' clearly and explicitly from scratch. This can be seen to be a deficiency in the traditional systematic approach, which does not explicitly deal with the importance of the coverage of issues and their interactions and, at the very least, does not emphasize the importance of these interactions in:

(a) developing a robust understanding of the issues that need to be considered within the process in order to fully address the nature of the problem and solution;

(b) exposing the true behaviour that would be achieved when a proposed solution is introduced into the operating environment; such behaviour includes the desired effectiveness that would be achieved and, equally crucially, the undesired 'side-effects' – in a true systems approach the focus of the evaluation within the process would be on exposing the 'emergent properties' (good and bad, desirable and undesirable) of the system in the real world.

In the following section we discuss the fundamental issues that must be established in order to develop a picture of the systems issues needed to achieve a robust systems approach.

3.5 Dealing successfully with systems

In the preceding sections we described the importance of establishing a robust understanding of the systemic elements, issues, and behaviours that must be addressed within the systematic process. So, what does it mean to 'understand' a system? We can identify the key aspects that must be established, by looking at the nature of systems themselves. Figure 3.3 illustrates the principal aspects of a system.

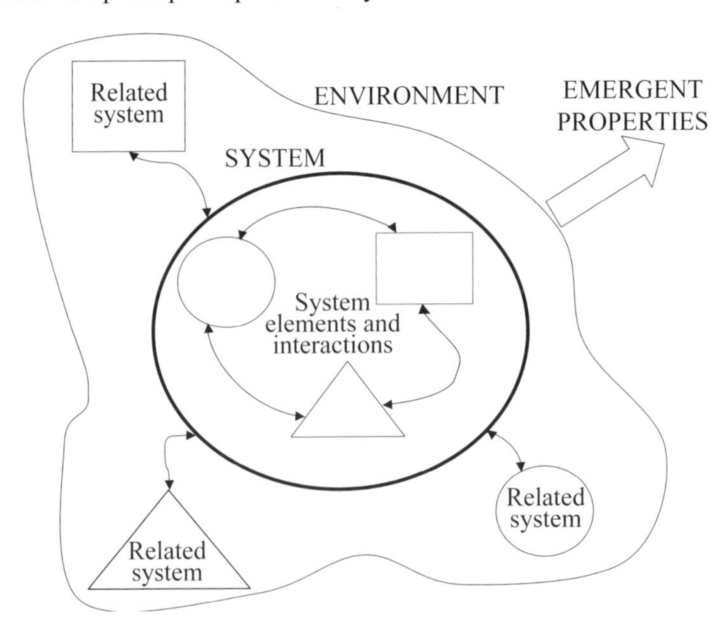

Fig. 3.3 Understanding systems

- We need to be clear about what is the system we are fundamentally interested in (the 'system of interest') – its elements and the interactions between them. What is inside and what is outside the system – what is the 'system boundary'? In many ways, we can choose this to be whatever we believe to be sensible or pragmatic. For example, in terms of an engineering design process, we could define this to represent the extent of our design responsibilities, that is, we can design (decide) things that are inside the system of interest, but we have to fit in with things that are outside.

- Related systems are those external systems in the wider environment that interact with our chosen system of interest. We do not 'own' these related systems and so we cannot impose design decisions on them – although we may attempt to negotiate some collaborative arrangement with them.
- Our system of interest and the related systems interact within a wider environment and it is from this total picture that the emergent properties emerge, since they are a property of all the elements involved and the interactions between them.
- In addition, in order to achieve a solution that is robust in the real world, we must also assess the degree to which all these issues and influences will, or may, change over time. In this way we can assess the ability of our proposed solution to adapt, or be adapted, to cope with changes in these aspects. This is, of course, related to the obsolescence problems with which engineers are all too familiar.

These general aspects of 'understanding systems' provide the basis for us to deal with the increasingly systems-oriented world in which we have to engineer effective solutions to real-world problems. It is obvious, even in the abstract description we have taken so far, that any deficiencies, inaccuracies, and misunderstandings in these fundamental systems issues would be very likely to introduce risks and problems into an engineering design process that had to deal with anything other than the simplest of situations.

3.6 Achieving a systems engineering approach in engineering design

We have now established the true fundamental building blocks of a systems engineering approach:

- A disciplined, systematic process that ensures that all decisions are visible, traceable, and justifiable, with a clear problem statement, solution options that are evaluated explicitly against 'needs' rather than judged on their features, and which takes an iterative approach to complex decision-making.

- A basis for establishing a robust understanding of the systems issues and influences that are relevant to the problem and to the solution, the product and the process; this must be addressed explicitly as we progress through the process.

So what does this mean in terms of achieving a systems engineering approach in engineering design? Well, the first crucial point to make is that engineering design must not be merely about the application of a systematic 'common sense' process, since this provides no basis for ensuring that our understanding of the situation and issues is appropriate. In other words, as with any other process, the old saying of 'rubbish in ... rubbish out' applies. This question arises even at the most basic level. For example, it is commonly recognized that the engineering design process must be based on multi-disciplinary teamwork and in very familiar, well-precedented situations. It is fairly clear which skills and expertise should be involved in such teams – in other words the 'system of interest' and the 'related systems' are obvious and well understood by all concerned. However, in a world of new technologies and processes, of rapid change and of complex, highly integrated environments, very few situations can be presumed to be well precedented with any confidence. Hence, the question of who should be involved in the multi-disciplinary team becomes less trivial. Perhaps the regulatory environment has tightened, or the customer is placing an increased importance on whole-life costs or on environmental issues related to disposal at the end of the useful life of the engineered product. All these issues would need to be represented in the decision-making process, and hence, potentially, in the multi-disciplinary team. So, there are more explicit decisions to be made relating to our reading of the situation and the issues that we judge need to be addressed within the process. This emphasises that the 'understanding systems' aspects must be at the forefront of our approach to engineering design in today's world.

3.6.1 Building a systems understanding

A true systems approach ensures that we develop a robust, common understanding of the problem – including our view of the environment and the issues and influences it raises for us to cope with, and similarly including our view of the potential solutions and the issues and influences that they raise for us to cope with. These systems views are, of course, not simple and one-dimensional. They raise complex, multi-faceted issues and interrelationships and it is highly unlikely that our first expression of the systems situation will think of all the aspects that need to be considered.

This is, of course, why the iterative approach of the systematic process is even more important in dealing with complex systems problems: we must not assume that our first statement of the problem and solution issues is correct. We must explore, discuss, and consult with a wide range of people with differing experience and domain knowledge, and at first there is likely to be rapid development of our system picture, as we add things that we had not thought of. Gradually, as we undertake additional, rapid iterations, we shall see the picture start to stabilize, indicating that we are approaching a more robust understanding of the systems issues.

3.6.2 A holistic approach

In building up a systems understanding we must take a holistic view, since there are many aspects to consider. We must recognize that the real-world solution that is sought, and the related systems and environment with which it interacts, are not just composed of obvious engineering design products such as hardware and software. While these engineered aspects are very important, in themselves they do not represent the full picture. What about the people involved in, or interacting with the system of interest? And the operating procedures by which any equipment is to be used? The full scope that we must consider for any system is:

- Hardware
- Software
- People
- Operating procedures

We can easily see that if any of these are deficient then the overall system is likely to be degraded – a clear indication that all these issues must be addressed throughout as part of the total system needed to achieve the real-world objectives.

Similarly, we must ensure that we take a complete view of the issues that (may be) important to our system and its successful operation. In particular, we must ensure that we address all the stages and situations in which the system may find itself. These include:

- Development
- Production
- Integration
- Installation
- Storage
- Transportation
- Use
- Training

- Maintenance
- Upgrade
- Disposal.

In considering all such situations, we shall be developing a much more complete and robust view of the effectiveness and limitations of any proposed solution, and hence informing the decision process, trade-off judgements, and so on – so achieving an effective solution with more confidence.

3.6.3 Evaluating decisions in systems terms

Earlier, we said that any proposed solution to the real-world problem must be evaluated in systems terms in order to expose the emergent properties (desirable and undesirable) that would arise if it were to be introduced. This evaluation is, of course, non-trivial for systems questions and must be conducted with full consideration of the systems relationships, as reflected in the systems understanding we have taken great care to develop. It is a feature of systems that their properties arise, not just from the characteristics of the elements that comprise the system, but also from the interactions between them, and between the system, related systems, and the environment. Hence, the behaviour of a system must be considered to be a dynamic thing, which cannot be evaluated by merely 'looking' at the design in a static way – it must be 'animated' so as to expose the emergent properties. Therefore, in a systems engineering approach, methods and tools that enable one to expose and explore dynamic, emergent properties of the system are absolutely necessary. For this reason the use of dynamic simulation and modern techniques such as synthetic environments are very important in a true systems engineering approach.

3.7 Implications for the engineering profession

In the preceding sections, we have described the importance of taking a proper systems approach to engineering design in today's complex, changing world. This raises some interesting questions on the implications for the traditional engineering domains, which are, of course, focused on their respective bodies of knowledge and expertise – indeed they are the guardians of these important capabilities. It is essential that the focus on the engineering domains is not lost, since it is from work at the forefront of the traditional domains that major technological opportunities and developments arise. However, we can see that engineers of all domains will find that their working environment in, say, industry is increasingly based on multi-disciplinary teamworking and a systems engineering approach. How should they be prepared for such an engineering environment, while still achieving their expertise in their fundamental domain of, for example, electronics or mechanical engineering?

3.7.1 Education

This clearly raises fundamental questions about engineering education and continuing professional development. It is crucial that all engineers are able to relate to the 'systems issues' discussed above, and to contribute to, and communicate through, the system views. It is our view that the fundamentals of a systems approach should be taught in every undergraduate and postgraduate engineering degree course, and possibly even introduced into the school curriculum. This may surprise some people since there is an opinion that one can understand the issues raised by a systems approach only if one has gained sufficient experience of real-world issues through, say, several years' working in industry following graduation. There is some merit in this view, but it relies on the individual to gain the maximum from his or her experience in industry. How much more productive and useful as a learning opportunity would the industrial experience be, if the engineer were prepared beforehand to look out for systems issues, and to relate the issues to each other, so building up a real-world systems view much earlier and more easily. There is an old saying that contrasts two people who have each worked in industry for ten years: one of them has ten years' experience while the other has one year's experience ten times. The first has been able to 'join up' his or her experiences, relating the issues to each other, to a wider context, and to fundamental underlying principles, while the second has taken them at face value and has gained no appreciation of the wider picture. We believe that engineering education, at all levels, should include a significant element of systems engineering which, even if done prior to the engineer gaining real-world experience, would provide a systems framework through which engineers can 'synthesize' their experiences and so gain the maximum benefit from them, accelerating their professional development in the process.

The professional engineering institutions are beginning to take an increased interest in systems engineering, as illustrated by the initiation of a 'Professional Network in Systems Engineering' by the Institution of Electrical Engineers. It is hoped that such initiatives will stimulate debate about the relationship between the bodies of knowledge of the traditional engineering domains and that of systems engineering, particularly in view of the fact that professional engineers now spend their careers invariably working with multi-disciplinary, systems-oriented processes and problems.

3.8 Summary

In this chapter we have discussed the fundamental principles of the engineering design process and of the systems engineering process, and have shown that they are essentially the same, in process terms. This emphasizes the fact that both of these processes are, traditionally, expressed as systematic, disciplined processes. We have also discussed the nature of systems and the increasingly systemic aspects of the complex, joined-up world in which we live, work – and engineer. The increasing complexity of the modern environment demands that we base our engineering approach not just on a disciplined systematic process – for if we do we risk the problems of 'rubbish in ... rubbish out' – but also on a robust understanding of the systemic relationships and behaviours of the complex systems we now deal with.

Hence a modern systems engineering approach must be based on two fundamental principles: a disciplined, systematic process and a robust systemic understanding of the challenges and issues to be tackled. This represents the environment in which today's professional engineers work and this trend is becoming stronger very rapidly. We recommend that all engineering education should include a significant element that prepares engineers for such a systems world; this would bring many benefits, including the acceleration of their 'joined-up' professional experience.

3.9 References

(1) **Hawkes, B.** and **Abinett, R.** (1984) *The Engineering Design Process*, Pitman.

(2) International Council on Systems Engineering, *Pragmatic Principles of Systems Engineering*, see www.incose.org.

(3) International Council on Systems Engineering, *2000 Systems Engineering Handbook*, Version 2, July 2000, see www.incose.org.

(4) **Stevens, R., Brook, P., Jackson, K.,** and **Arnold, S.** (1998) *Systems Engineering, Coping with Complexity*, Prentice Hall.

Chapter 4

The Enterprise Environment for Engineering Design

4.1 Introduction

Engineering design is, of course, not an isolated activity. It is always conducted within the context of a wider environment and this has a fundamental influence on the expectations and objectives that the design must satisfy, the constraints and opportunities with which it is faced, and hence on the conduct of the engineering design approach itself. This environment has changed out of all recognition over the past century, and it is important to understand these changes and the influences they have had on engineering design and the products that are produced. If we do not understand the context in which engineering design was (and is) conducted then we shall not appreciate the engineering decisions that were taken, the judgements that were made, and the reasoning behind them, and we will not 'learn from experience' effectively. Learning from experience has always been important in engineering but it must be appropriate and relevant. Merely copying previous design decisions and engineering approaches without judging whether they are appropriate to the current situation is risking failure and, at the very least, is likely to stifle innovation. This is particularly true when change is rapid and, of course, change has accelerated in the last 10 or 20 years, with the 'information age' affecting all aspects of our lives, including the technologies and organizations that shape the environment for engineering design.

In this chapter we explore some of the features of this 'enterprise environment' and look at how they have changed, particularly over the last

100 years. We choose, in effect, to look at the changes that occurred during the twentieth century, since during that time engineering design 'came of age' as a formal, recognized discipline. It is interesting to address the following questions and compare how they have changed over the years, including the implications of these changes on engineering design:

- What is the environment in which engineering design is undertaken?
- What expectations does the 'enterprise' place on engineering design and its products?
- What are the criteria by which success is judged for engineering design?
- Who makes the engineering design decisions?
- What do they deal with in terms of technologies and expertise?
- How do they organize their engineering design approach, who is involved, and when and how are ideas and decisions communicated?

4.2 Early engineering design

Engineering design and its products have, of course, been an essential part of human activity and progress for millennia. One has only to look at the products of the architects, masons, and shipbuilders of ancient times to acknowledge their design skill and expertise. These people were the ancestors of today's professional engineers, with such civil and military products as buildings, ships, and weapons being engineered long before the Industrial Revolution. However, the engineering design these skilled people undertook was very much based on their 'craftsmanship', passed down through the generations from master to apprentice. Until fairly recently, engineering design was not the formalized discipline we identify today. The professional engineering institutions that represent engineering as a formal profession today were founded within the last two centuries. For example: the Institution of Civil Engineers (1818), the Royal Institution of Naval Architects (1860), the Institution of Mechanical Engineers (1862) and the Institution of Electrical Engineers (1871). In fact, through most of history engineering design has been very much a 'workshop' sort of discipline, conducted in a very intimate way by people who were 'masters' of their domain, and there are many examples of ingenious designs being devised to solve real-world problems. There were, of course, far fewer technologies and processes available to them than there are for today's engineers and they could develop a real expertise in them, built on centuries of experience and the evolution of designs, processes, and craftsmanship through experience. Their focus was also very practical from the outset, and these craftsmen were closely associated with the manufacturing of the products as well as their design. Consequently there was a very intimate relationship between design and manufacturing skills.

As an illustration of the environment for engineering design in the past, consider the following view of a product being designed. Some 'master craftsman' has identified a real-world challenge and has a vision in his head of the product to be produced to solve it. He has some objectives to meet and is the master of the design process and maintains close control of it throughout. He liaises with experts in individual technologies and processes in a very intimate way: imagine, say, the Wright Brothers sketching their vision of a powered flying machine and discussing it with carpenters and mechanics. In fact, they undertook much of the work themselves, using their own expertise in the technologies being applied, and learning from their own experiments and experiences. In early engineering design the engineering designer is very much the visionary for the design. Also, the vision of what he is seeking to achieve is dominated by the immediate nature of the end product – its 'performance'. For example, the Wright Brothers were seeking to design and build a powered aircraft that could carry the pilot for some distance, and success was clearly demonstrated on the day of the very first flight. That first powered flight was the culmination of many experiments with gliders and prototype designs over a number of years.

The approaches that characterize such 'early' design engineering may be thought of as:

- A single person holds the design vision in his head (or in the case of the Wright Brothers, a close-knit pair of people).
- He works very closely with appropriate craftsmen in what we may think of as a 'workshop'-type enterprise.
- The 'chief designer' produces fundamental sketches and drawings of his vision in order to communicate them to others.
- The workshop mode enables expertise to be highly integrated into the decision process. (In fact, since there were relatively few technologies available to the designer, he was often a master of them.)
- The objective is to produce a 'working product' in accordance with the vision.

We can see that the environment in which early engineering design is undertaken is very intimate. The design is fairly well contained within a single person's vision and, while the manufacturing effort may be on a very large scale (such as with ships), the design process itself is focused on a single individual. What about the expectations placed by the 'enterprise' on engineering design and its products? There have always been cost constraints for engineering design to consider, of course, and these would have to be estimated and dealt with by the engineering designer as part of his 'craft'. However, these considerations would have related to relatively immediate concerns, such as the amount of funding needed to build the product. Other immediate criteria would relate to the performance of the product itself. So, the enterprise expectations were relatively straightforward and direct – get it working and do not spend more than 'X'. Success was judged by meeting these criteria, with the performance and the achievement of an effective product being dominant.

The design visionary who made the central design decisions dealt with a limited number of technologies and processes and was able to understand the important issues that influence the design decisions from the outset. It was clear which skills and people needed to be involved and consulted in the decision process, and for many issues the engineering designer had expertise in them himself. The conduct of engineering design was largely informal, with the designer communicating his decisions through drawings and discussions with appropriate experts in a very close and direct way, in keeping with the 'workshop' environment in which they worked.

4.3 Formalized (system) engineering design

Earlier we mentioned the formation of the engineering institutions, each being focused on a particular facet of engineering, such as 'mechanical' or 'electrical', and usually associated naturally with a particular range of technologies. This was driven by the expansion of a body of knowledge to underpin engineering design. The industrial revolution and the discovery of new scientific and technological principles, with a more scientific approach to the understanding of physical characteristics, led to an explosion in the scale and sophistication of the body of knowledge available to engineering designers. The institutions sought to capture this and formalize it so that a more rigorous, scientific approach was taken to engineering. Engineering designers were able to draw on, and contribute to, this body of knowledge but the process of engineering design itself was still dominated by the strong individuals and workshop-like environment described above. One can identify this approach in such eminent designers as Ford, Barnes-Wallis, Mitchell, and others who had design vision. Henry Ford is, of course, famous for implementing revolutionary manufacturing practices, but the underlying engineering design of, say, the Model-T Ford was still dominated by his vision, and he was able to discuss a wide range of technological issues with craftsmen and experts in a range of technologies and processes.

As the twentieth century progressed, the number and diversity of technologies available to an engineering designer increased dramatically, particularly with the impetus of the Second World War (for example, jet engines, radar, electronics, computers, and so on). Designs also increased in scale and complexity. As the new technologies developed it became impossible for one person to keep track of the developments, and engineers became increasingly specialized. The design became 'more than a head-full' and so did the body of knowledge associated with the increasing number of technologies used in it. So, on the one hand engineered systems were becoming more and more multi-dimensional and diverse in terms of the range of technologies they contained, and on the other, engineers became more specialized. This presented the engineering design approach with a new sort of problem. Whereas previous design had been conducted with a single visionary and with intimate 'workshop' enterprise relationships, now engineering design called for much more sophisticated liaison and decision-making across and between the multitude of engineering specializations. Further, the interdependencies between their engineering decisions were more wide-ranging and complex than ever.

So the latter half of the twentieth century saw a significant change in the scale and complexity of engineering design that drove it away from the previous workshop-type enterprise environment. Engineering design now had to be co-ordinated across and communicated to a wide range of disciplines, each with a specialized, sophisticated body of knowledge that was not fully understood by others. In addition, larger teams were more dispersed geographically. This led to a need to communicate detailed design intentions much more explicitly and formally than previously. A crucial aspect of this was the identification and communication of the interdependencies and relationships between design decisions in the various contributing areas: the engineered product must be thought of as a 'system' of interacting, co-operating parts driven by the achievement of an overall objective and with system-level design decisions on what each contributing part needed to fulfil. In short, as engineered products became larger in scale and comprised more diverse technologies, engineering design needed to be unified through the engineering of the overall system.

It was in response to this changing situation that formalized engineering design was developed, with formal communication of the design objective, the elements comprising the design, and their interactions. This formalization of 'systems engineering' happened in the years following the Second World War. The approaches that characterize such 'formal' system engineering may be thought of in the following ways:

- A single person can no longer hold the design vision in his head.
- There are many more technological specialisms and their role and relationships within the design must be co-ordinated.

- The scale and diversity of the technologies means that as engineering design was outgrown a workshop approach and the communication has become less intimate and more formally documented.
- Formal requirements and interface control documents are produced in order to inform a specialism what is needed of their local design.
- These documents are of course written as paper documents and drawings – a relatively slow medium – and so the engineering design process tends to be sequential in nature; much more so than in the workshop environment of the early twentieth century.
- The objective is still to achieve a 'working product' in accordance with the vision.

We can see that the situation in which this engineering design is undertaken is very different from the intimate, workshop environment. The design is now so large and diverse and the body of knowledge so extensive and specialized that it can no longer be held within a single person's expertise – more formal, sophisticated means of communication, discussion, and decision-making are applied. But what about the expectations placed by the 'enterprise' on engineering design and its products? In fact these enterprise expectations are still relatively straightforward and direct – get it working and do not spend more than X. The focus is still on achieving the immediate engineered product, such as a heavy bomber capable of carrying a certain payload a certain distance at a certain speed. These enterprise expectations drive the basis on which engineering decisions are made and so pervade the whole process, with formal specifications being focused on performance, cost, etc. However, these expectations have changed significantly, particularly over the last couple of decades of the twentieth century and they have a fundamental impact on the conduct of engineering design.

4.4 Enterprise expectations

In the preceding sections we described the objective that was driving design as being dominated by the achievement of a working product in accordance with the vision. This is, of course, a simple statement and in many cases there would have been some sort of limitation on available cost and time. These are examples of stakeholder expectations and constraints. In addition to the obvious technical objective for the designed product, any engineering design effort will be influenced by some stakeholder expectations and constraints, and these, of course, influence the approach that is taken.

In the early part of the twentieth century the enterprise objectives were dominated by the direct technical achievement of the designed product (such as powered flight), and the design effort was relatively informal and focused very much on this aim. When system engineering was first formalized following the Second World War, the focus was still very much the same as far as design was concerned. Of course, over the preceding years larger industrial enterprises, such as Ford, had grown up but these did not have a great influence on the design environment itself. Henry Ford provides a good example of this: his innovations were predominantly in the production area, with factories becoming highly organized and geared up for mass production.

In the last couple of decades of the twentieth century, enterprises became much more interested in meeting highly competitive time and cost constraints, both in manufacture and in development, and the interest in and expectations on non-performance issues increased significantly. The previous focus on achieving the technical product was augmented by an equal desire to reduce time and cost overall. Cost of development and production, time to market, etc. became major enterprise expectations and it was realized that a major problem contributing to these issues was that of rework in production caused by the engineering design needing to be 'productionized' following its communication to manufacturing. Further, it was realized that the most cost-effective time to address these issues was during the design phase so that ease of production and assembly was enabled by better design. This showed itself in the increased focus on 'design for manufacture' and the development of approaches such as concurrent engineering, aimed at improving the communication between specialist areas, particularly as design related to production and achieving improved decision-making early in the process. The more extensive and sophisticated enterprise expectations led to significant changes in the engineering design approach that was followed; its principal features were:

- Improved decision-making as early in the design process as possible in order to reduce expensive, time-consuming rework and changes late in the life cycle (especially in production).
- Improved communication between specialist areas, including the formation of integrated development teams containing multiple disciplines and, in particular, engineering disciplines and production.
- Introduction of computer-aided tools to reduce time and ease communication. In particular, computer-aided design and manufacture computer packages enabled design information to be electronically provided to production without the information having to be significantly reconstructed.
- Increased use of techniques that enable the implications of proposed

decisions to be investigated early in the process. These include rapid prototyping, simulation, and modelling.

In this enterprise environment for engineering design, there is a great focus on effective decision-making, enabled by close, multi-disciplinary teamworking and sharing of information. In many ways this is trying to recreate the intimate environment of the 'workshop' era, but in a situation of much larger-scale systems, more diverse, specialized technologies and processes, and – crucially – with the overall design vision having to be progressed by a team of people rather than by one visionary.

Originally this trend was driven by enterprise expectations of reducing time and cost and improving the management of risk in the development and production of engineered products. However, in recent years there has been increasing emphasis on achieving wider success criteria. Enterprise expectations have grown to encompass issues that are wider than production. The effective use of the product by the end-user/consumer has become an increasingly significant enterprise issue, not least because issues such as reliability, low-cost maintenance, and ease of use are seen as important cost-reduction areas and marketing features. Hence the design approaches that had developed as concurrent engineering, which originally started with a design-for-manufacture focus, have now expanded to include a much wider view of the life cycle. This has drawn even more disciplines into the early decision process, including maintainers, upgraders, transporters, and of course users.

The trend continues, as seen by the increased interest in environmental affairs, leading to much more focus on design for disposal and/or recycling in, for example, automobile design. If enterprises were still interested only in achieving the immediate performance of an engineered product then many of these changes in our approach to engineering design would not have occurred. These enterprise expectations demand appropriate methods and tools to be employed in engineering design and systems engineering to achieve the aims. Hence, the scope of rapid prototyping, simulation, and modelling is expanding to encompass the range of issues relevant to the enterprise expectations. This demands the development of (valid) models of the manufacturing process, the logistics chain, the environment with which the users and maintainers are presented, and so on. So, engineering design can meet these challenges only by developing means of addressing the questions raised.

Over the last decade or so there has been another significant change in the enterprise environment for engineering design. The organizations themselves have become much more global, and members of engineering design teams can now be distributed literally anywhere in the world. Indeed, some industries take advantage of this by achieving 24-hour working, again contributing to reductions in time. Such a situation is an extreme example of distributed teamworking, but the trend is apparent to an increasing degree in almost every industry and engineering sector. Since information must be communicated across the team in an integrated, clear, and timely manner, the distances between them must be overcome and, of course, modern information and communications technologies have been a crucial enabler in making this possible. Hence, we see a great emphasis in modern engineering enterprises on infrastructure investments, such as shared data environments (SDE), knowledge management, integrated toolsets, video conferencing, etc., in order to achieve a responsive, integrated engineering design approach across large, diverse teams, even when they are separated by large distances.

4.5 The nature of engineering design in modern enterprises

Design in engineering may be thought of as effectively encompassing three aspects:

- Conceptual design
- Embodiment design
- Detail design.

Most people, when they talk of engineering design, tend to think of embodiment design, that is, where the concept is turned into shape and form. Detail design finishes the process with the details that transform shape and form into meaningful dimensioned shapes that can be manufactured. Embodiment and detail design are the domain of CAD (computer-aided design), and the introduction of this system over the last 25 years has revolutionized design office practices, but more importantly it has revolutionized design office capabilities. Much of what we have described above as some of the elements of systems engineering would be very difficult, if not impossible, in the old-fashioned drawing office. Interaction in the design process is fundamental to systems engineering. We might consider whether the historical, evolutionary path in engineering design, described in an earlier chapter, was not partly a product of the way design offices were laid out, when separate design sections were tasked (by the principal design 'visionary') with the design of particular sub-systems. Sub-systems integration as an explicit design issue was not a main feature until quite late in the evolutionary trail.

It is important to realize other important facets of CAD, which go largely unnoticed. First, it has eliminated the need for the designer to be a competent draughtsman as regards his drawing skills, with the consequent long training period this always entailed. Second, it has produced more uniformity in design, particularly through the use of standard subset routines, which eliminate different personal interpretations of drawing standards and so on. Thirdly, and increasingly, it has provided the necessary platform for the formation of virtual design studios (VDS), which allow design teams to be assembled rapidly from the best (or the most appropriate) experts in the world or from within the company, despite geographical separation and time differences. The conventional meetings between people involved in particular activities will always have a place, due to the human contact involved, but in an era of globalization, the virtual meeting is becoming ever more important. Downsizing of mainstream engineering companies, and the consequent reduction in their design departments and increasing use of outside consultants, all suggest that impromptu virtual design studios will increasingly become the design norm in the future. This is only really possible through the use of CAD and related systems. At the same time, there are a number of fundamental problems to be researched in VDS before the full potential can be realized. The flying around the world (or even around Europe or America) of particular design experts or specialists, for two- to three-hour face-to-face meetings, does appear ever harder to justify, however. VDS is the answer, but it still needs development, not only on a basic level but also to suit particular local, company, organizational and cultural conditions. Allied to VDS is the virtual reality process, which in particular is useful when coupled to rapid prototyping techniques.

These techniques enable the what-if syndrome to be exploited throughout the design stage to significant advantage, and recent rapid prototyping advances mean that this can be further developed to provide useful component shapes, capable of being tested. These developments all contribute towards ensuring the design is correct before significant project milestones have to be met.

The use of finite element analysis (FEA) and computational fluid dynamics (CFD) is, of course, now a standard procedure in many aspects of engineering design. FEA and, to a lesser extent, CFD have become mainstays of engineering degree courses and in mainstream design practice, but the validity of the simulations used for a particular purpose – to aid a particular decision – must always be very carefully assessed. This is particularly true here since much of the derived information produced by these techniques is not easily rechecked. In some respects the use today of computers and even calculators implies a degree of accuracy in the derived data which may be unjustified. Many designers will recognize issues such as that illustrated by the experiment in a fluids laboratory where a fluid was being weighed in imperial units. When the technician was asked to give the weights in decimals the 5 lb 7 oz weight became a mathematically correct 5.4375 lb, but the accuracy of weighing certainly did not justify an answer to four decimal places. Similarly, it is all too easy to assume accuracies in our derived figures that are not actually there.

Conceptual design is a more difficult phase to mechanize by CAD systems and their like, although in many respects it is the most important, and certainly the most innovative, phase of the total design process. There seems to have been relatively little research carried out on the conceptual design phase, which is the phase in the design process that is probably much more centred on the 'scientific art' aspect of engineering than the other two phases. It is at the conceptual design phase that many of the basic decisions necessary have to be made in the systems engineering approach to design. It is here that a design strategy called strategic design can be used. It is a basic principle of the systems engineering approach to design that all aspects and inputs are considered at the initial conceptual stage.

4.6 Strategic design

Strategic design is an activity concerned with creating and operating a coherent long-term design strategy where the core activity is an ongoing selection and evaluation of concepts (primarily those that determine product technologies and architectures). It is concerned with searching for solutions that offer the best prospects to meet short- and long-term customer requirements (and expectations) as closely as possible. The aim is to avoid running into commercial and/or technological dead ends as market conditions and technologies rapidly evolve. Strategic design does not restrict design solutions to areas where known solutions currently exist. It looks at trends and strategies in the market place to identify where long-term opportunities will exist, even if they are outside the scope of current design capabilities. As discussed earlier, in the past so much of design thinking has been driven by short-term enterprise requirements, with little regard to the middle- and long-term needs. This way of thinking is still apparent in most of Western industry, in spite of the intention to take wider life cycle considerations into account, and it exists at the highest level, despite many case studies quoted at management schools! Strategic design then is an integral part of systems engineering. Long-term *requirements* can often be specified with some precision, even when design solutions are not yet available.

The writer developed this 'strategic design' strategy in America during a period directing design in a manufacturing company producing machinery for the corrugated board industry. A new range of large and complex machines to handle and form paper products was to be designed. One approach was to modify the existing design for a quick solution, but this would produce a range with a short market life. Instead the likely requirements of the customers over the next 10 to 15 years was predicted. By looking at customers' requirements in detail it could be discerned that a continuing middle- to long-term trend was a move towards ever-increasing output quality, using ever-poorer-quality product materials, particularly the use of recycled paper. A method of improving output quality was identified, although at that time the sensors, transducers, and actuators that would be required were not available and those that were available had insufficient capabilities. However, the strategy laid out a long-term plan that enabled particular machine capabilities to be incorporated when the technologies required were sufficiently developed. Of course the strategy also identified the need to pressurize the appropriate suppliers to develop their sensors and actuators to meet the requirements at particular points in time. Marketing requirements to produce new versions of the machines evolved and the technical specifications needed at that time were also then apparent. The strategy was very successful and led to a new range of products with a long-term as well as short- and middle-term marketing programme.

While this is not easy, it is generally possible to see trends and extrapolate to likely scenarios. In today's world, an increasing requirement for flexibility, quality, consistency, and throughput are discernible. While the means of achieving these requirements are not always technically available yet, they can be set as requirements for the design over a time-span. The designers are then able to offer the marketing department machine design where a given standard can be 'pulled out' at whatever time suits the market place, without the usual panic reaction to a new requirement. If a long-term, fundamental view is taken of all the considerations that have to be addressed, particularly the long-term effects of particular design decisions, then a rational design plan can be established. This technique tries to avoid the familiar scenario of designs that have a limited life due to obsolescence of choices made early in the product's life cycle. It seems very obvious when set down, but too few managers seem able to consider anything other than the short-term solutions.

4.7 Simulation

A key guide in helping us develop a strategic design approach is the use of the simulation and prototyping techniques mentioned above. As our systems become more and more complex, and their interactions more difficult to predict, we need to look at what aids there are to help us understand the complete whole, its internal interactions, and its external outputs and influences. Perhaps the most powerful aid we have is simulation techniques. With accurate simulation much guesswork and actual system testing can be minimized, and increasing use is being made of simulation techniques to predict performance, particularly in very complex systems. The absolute cardinal rule of simulation (and in very many other areas that we use as well), is that the final accuracy of our simulation is dependent on both the accuracy of the initial information and the assumptions that we make and put into the simulation. 'Measuring with 0.001 per cent accuracy and 10 per cent validity' is a saying ascribed to Professor Stein at Arizona State University that is well worth memorizing, since it encapsulates this very problem. It is also expressed as 'garbage in, garbage out', but Professor Stein's statement is more explicit, and certainly more elegant! Engineering designers and systems engineers should also remember the old adage 'Every model is wrong ... but some of them are useful'.

Simulation enables us to see how the design will respond to various situations, and in appropriate cases a range of different circumstances should be animated, to enhance our understanding of potential performance. To take a mechanical example, many mechanical design simulations are based on the use of FEA, which can be thought of as a computational extension of physics. Generally, the finite element method is used where the shapes of objects are so irregular or complex that finding simple formulae for their boundaries is intractable. With the representation based on regular polyhedrals, complex shapes become computationally tractable because the physics can be broken down over the polyhedrals and then reassembled over their faces in coherent ways. In static and semi-static designs (such as buildings, cranes, ships, and the like), simulation can be allied to CFD analysis to check air-flows, interference patterns, and so on. In machinery design, component interference patterns can again be checked before sizes are finalized for inertia analysis. The system design characteristics in a real-world scenario can thus be estimated at an early stage in the design process.

In our view, simulation will take an even more important role in design in the future. The problem that is then facing us is in deciding whether we have modelled in sufficient depth in our simulation to cover all real-world conditions. Have we covered every failure mode, every failure situation? Have we built into the simulation a feature that, in some crucial aspect, does not fully reflect how things would actually happen in the real world? For instance, consider the case of a light aircraft with a control malfunction, flying towards a cliff face. The airframe design is stressed to, say, 5g but avoiding the cliff face necessitates a 6g turn. Do we design the aircraft control system to restrict any turn to 5g, or do we allow the pilot to attempt a 6g turn? (The manufacturing differences between one airframe and another mean that the 5g is a minimum figure, but might be capable of being exceeded.) In any case, is not the normal human reaction (in this case, of the pilot) to attempt to avoid certain death, even when the consequences of such action are not known? We should put these considerations into our simulation, but how do we define them with sufficient accuracy? How do we know if we have modelled enough – every failure mode? Every situation? Can we still achieve the level of confidence we are used to?

Generally, complex systems have to be modelled and analysed within computers. Representation is then one of the central problems: how can objects and activities that seem so obvious to human beings be represented inside computers? Physics has developed the mathematics of space–time as an incredibly powerful way of representing objects and their dynamic interactions. However, even this representation has limitations, because systems usually have to be represented at different levels of detail and aggregation, corresponding to our methods of observation. In particular, when systems are computationally irreducible, any attempt to predict their behaviour is, effectively, a simulation. Some of the constraints are due to the nature of parallel-distributed computation. Of more relevance is the 'Can you trust it?' problem of simulation science. A computer simulation involves the mapping of a real system on to explicit objects or structures within computers. The computer then executes transition rules that compute new system states from old. By running the simulation through time, system states can be computed into the future.

The problem lies in the appropriateness of the mappings on to the representation, and the correctness of the transition rules. There are many examples of simulations in which the representation is clearly inadequate and thus the transition rules do not adequately reflect reality. Would we trust such a simulation? At present, we have no simple test to check the correctness of a simulation. Of course, there are many considerations that can be invoked to give degrees of confidence, perhaps simplest is the 'Popper Test': if the simulation gives results inconsistent with our observations, then the simulation is most probably incorrect. This sort of test is acceptable if we are able to make deterministic predictions and match them with equally controlled measurements from real life. However, if the system is chaotic, a single observation or a single simulation is inadequate. In this case, one needs to make many observations and work within the statistics of distribution of outcomes – if one is able to determine the likely statistical behaviour.

Another difficulty is where the problem involves very many variables, such as an analysis of the behaviour of the blades and stators in an axial compressor, as commonly used in aero gas turbines. Because of the number of variables and the complexity of their aerodynamic interactions, we need to look to complexity science and adopt a self-organization approach as distinct from a top-down control. In complexity science, it is common to use autonomous intelligent agents and use this approach to create a self-learning control architecture.

There are many examples of self-learning systems, and we take part personally in many, on every day of the week. The flow of traffic through a town or city, the crowd at a football match, and pedestrians in a shopping precinct all demonstrate this type of behaviour. It is common in nature, too, where examples include birds and bees that self-organize into flocks and swarms. As far as we are aware, there is no chief bird or chief bee directing operations: pure systems behaviour emerges from the actions and the interactions of the individuals. This type of behaviour is becoming increasingly important in our designing of control and, where appropriate, needs to be taken into account in our simulation studies. If our systems become truly self-learning, can we then still achieve the level of confidence in our simulations that we are used to? It would seem difficult to put the exact same self-learning capability into a simulation technique.

4.8 Summary

So, we can see that enterprise expectations have a direct, fundamental, and significant influence on the conduct of engineering design. From its original workshop-type enterprise the environment and expectations on engineering design became formalized and less intimate and then, in some ways, came full circle as whole-life issues necessitated the (re)creation of a more involved and intimate design decision process. Approaches such as concurrent engineering are seeking to recreate the intimate involvement and interaction between disciplines that existed naturally within a workshop environment but which had been compromised as the scale and diversity of technologies and systems expanded after the Second World War. However, while the underlying approach may at some level have some similarities with the intimate decision-making that was possible in a workshop, the reasons behind it are very different. In the early part of the century, the workshop was the natural place for such activity, since it grew out of the craftsman's environment, which had existed for centuries. In the latter part of the twentieth century, the joined-up decision process was being driven by hard enterprise objectives on cost and time and, moreover, it had to be achieved in much larger, even global, enterprise structures that have fundamental influences on the engineering design approach.

Chapter 5

Control and Embedded Intelligence

This chapter comprises mainly material contributed by Anthony Lucas-Smith.

5.1 The traditional view of control

Control is a long-established discipline that has been applied to industrial processes and machinery for many years. Bissell (**1**) defines control engineering as 'the design and implementation of automatic control systems to achieve specific objectives under given constraints'. Traditionally it has applied to well-defined, enclosed systems in which parameters such as pressure, temperature, linear and rotational speed, and position of an object or components are critical to operation. Typical examples include:

- A chemical production process controlled to operate at pressures and temperatures which are both safe and capable of providing optimal yields.
- An industrial robot that can pick up delicate components without crushing them and place them accurately in planned positions.
- A camera which can automatically adjust its lens position to achieve sharp focus on an object.

Techniques involve the creation of mathematical models to represent real systems as closely as possible. A typical strategy is that of closed-loop control in which parameters are monitored to observe if they diverge outside their specified limits. As divergence from the norm is observed, the necessary feedback, calculated from the model, is applied to adjust the parameter, constantly keeping it within the limits. The industrial robot referred to above might use a feedback loop to control the pressure exerted by its gripper. In some systems, feed-forward methods are used to control performance. In a lift in a building, the power applied to control its movement would be dependent on the loading of the lift. A feed-forward circuit would measure the load, calculate and then apply the power required to compensate for it, and move the lift smoothly.

5.2 Control as a more flexible system

Traditional control is well established and will continue to be essential for many processes and devices, particularly in an industrial, manufacturing context. However, the range of automation being designed and sold is in a state of great expansion. The application environment is widening, the range of parameters to be monitored is increasing, and the nature of the 'skills' required by automation is correspondingly more complex. Use of the term 'skills' implies a human or anthropomorphic connection and this is indeed a good way to characterize many current developments.

If we require a machine to 'see' a situation, recognize it, and then adjust its actions, then we have to give it capabilities that emulate human abilities. We use eyes and brain to capture an image (maybe moving or changing), interpret its relevant components, deduce the significant aspects of the situation, assemble a plan of action to be carried out, and then carry it out. We take it for granted that many humans can spot a ball being thrown towards them and then adjust their position to catch it, but replicating this behaviour in a machine requires state-of-the-art technology. The processes of seeing, interpreting, understanding, planning, adjusting position, feeding-back, timing actions, and knowing when the action is complete have all to be replicated in a machine. This might be seen as a rather extreme example, but many machines are now being required to operate in ways that closely mimic human capabilities, with the intention that they can do so at least as well, if not better, and more cheaply. In comparison with traditional control systems, such machines are required to operate in increasingly uncertain environments in which the range of possible situations and resulting actions is not easily defined. For example, a pick-and-place machine handling a defined range of components of uniform dimensions and weights, and placing them in well-defined locations, can operate successfully using a traditional control system. A mobile robot required to roam freely around a factory picking up loads of variable size and weight and deliver them accurately and safely requires considerably more advanced capabilities.

The widening scope of system boundary, in comparison with the environment of traditional control engineering, brings the need for more varied capabilities in sensing. Many of these are related to human abilities, such as seeing, touching, hearing, and smelling, and so require appropriately analogous sensors to eyes, fingers, ears, and noses.

Other sensors are used to recognize a range of physical phenomena, including:

- Positional (linear displacement, orientation of objects)
- Static mechanical (mass, pressure, torque)
- Dynamic mechanical (velocity, flow rate, acceleration, vibration rate)
- Electromagnetic/optical (wavelength, frequency, polarization, intensity)
- Physico-mechanical (viscosity, humidity)
- Electrostatic (charge, capacitance, inductance, resistance)
- Magnetic (flux)
- Thermal (temperature)
- Acoustic and ultrasonic (frequency, intensity, direction).

5.3 The perception–cognition–actuation model

We are familiar with the sensing/actuation process in human activities, to the extent that we mostly take it for granted as it happens. We see an object that we are looking for and pick it up. The link between sensing (seeing) and actuation (picking up the object) is, however, complex brain-based activity. Our visual system collects the light reflected and assembles it into an image that our brain interprets. Having made the association between the recognized object and the need to pick it up, the brain plans and controls the physical processes of moving body, arm, and fingers. The analogous situation in the operation of an intelligent machine is the use of sensors to provide data about some aspect of the real world, linked to actuators that can react physically with the real world. The linkage between sensing and actuation is expressed as two functions: perception – gathering information about the real world; and cognition – deciding, or planning, how to react with the real world. This is shown in Fig. 5.1.

The direct link between perception and actuation pushes the analogy with human activity a bit further, recognizing 'reactive behaviour', such as withdrawing one's fingers from a hot object without consciously planning the movement. In machine terms, this is equivalent to a certain phenomenon being sensed and an immediate, standard, actuation response being triggered.

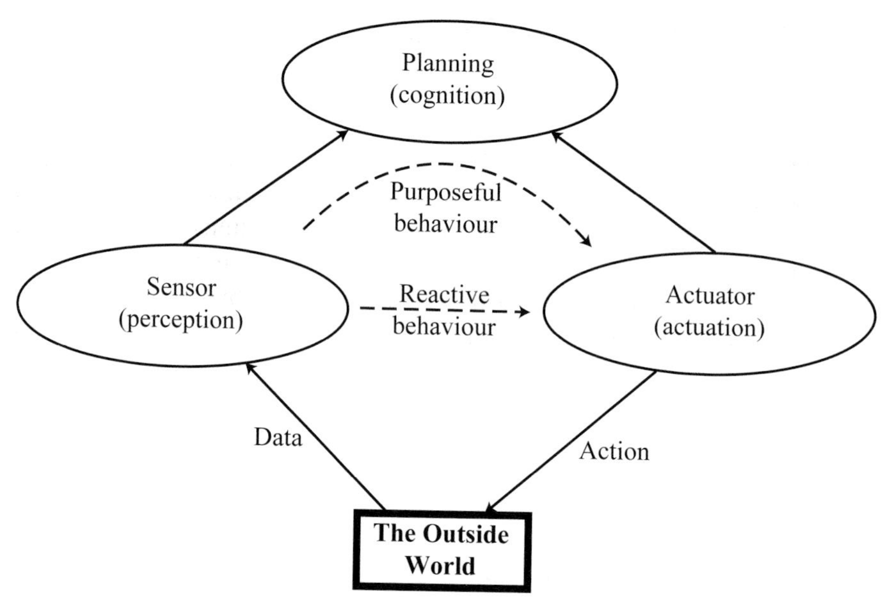

Fig. 5.1 Perception/cognition/actuation
(Adapted from Mechatronics: designing intelligent machines, 1995, edited by G. Rzeuski, Butterworth Heinemann/Open University.)

5.3.1 Sensor fusion

Just as humans use a combination of senses to make sense of the world, so machines can use a combination of sensors to obtain more detailed information about the state of the real world. We might investigate a substance or material through a combination of looking at it, touching it, and smelling it. Suspicion that an object is made of lead, from its appearance, might be confirmed by picking it up and appreciating its high density. Seeing yellow fumes produced by a chemical process might alert us to danger, confirmed as chlorine by a distinctive smell. Similarly, components of a machine might include two or more sensors, in order to:

- provide sufficient information to identify an object or a situation where information from a single sensor would be inconclusive, or;
- provide corroborating evidence where uncertainty is present.

For example, a vision sensor might identify the possible presence of an object, while the additional sensors of a co-ordinate measuring machine could be used to identify the object uniquely.

5.4 The need for embedded computing and communications

For many years the term computer referred to a mainframe with communication links to remote devices. With the advent of ever-cheaper processing power and memory, on a chip, computer power can be distributed and associated with a range of discrete functions. Examples include complex signal conditioning of sensor outputs, and microprocessor control of electric motors. 'Embedded computing' enables better decision-making through detailed knowledge of local conditions, but it also entails the need for extensive communication among such devices. Modern automobiles use information from sensors located in steering, braking, engine, and other sub-systems – all connected to a controller which returns signals to actuators in the sub-systems – to improve safety and comfort. The subsequent communications connectors can result in complex wiring, though the introduction of single-bus connection and radio frequency links should result in simpler communications in the future.

5.5 The nature of embedded computing and the concept of artificial intelligence

Intelligence in human beings can be interpreted in a variety of ways, related to: quick decision-making, fast recall of facts on many subjects, being good at carrying out difficult tasks, being very perceptive about others' behaviour, being intellectual, and others you may care to identify. To some extent these abilities have their parallels in machines which possess computational abilities, giving rise to the term artificial intelligence (AI). This is often loosely referred to as 'intelligence', somewhat supported by the anthropomorphic vocabulary used. AI systems may be said to decide, see, calculate, plan, assess, play, and so on – all words that describe human actions – but they do not carry out the same functions in the same way. A computer may be said to 'play' chess and therefore be (artificially) intelligent. What happens is that it has been programmed in such a way that given an input of a chess position, it outputs a chess move of a type that a human might specify. It therefore creates a good illusion of 'playing' chess and may be programmed well enough to 'beat' a 'competitor'; the quotation marks indicate how difficult it is to describe AI without recourse to human-centred vocabulary.

The distinction between AI and traditional computing appears to be rather arbitrary. We define AI here as the *capability of computational reasoning to pursue goals or desired behaviour while coping with uncertainty*. It could be argued that many traditional computer systems, such as programmable logic controllers (PLCs), are designed to handle uncertainty, in this case the uncertain combination of physical conditions in an enclosed process. We need therefore to look more closely at the nature and extent of uncertainty if we are to make a distinction.

5.6 Uncertainty and intelligence related to games and industrial processes

5.6.1 Noughts and crosses

Playing involves minimal uncertainty as there are at most nine positions where an O or X can be placed, and a game can only last from five to nine moves. A computer program to play the game needs no AI.

5.6.2 Chess

Playing is similar to noughts and crosses in that the movement of pieces follows strict rules and there are consequently, at any stage of play, a finite range of moves possible. However, it differs significantly from noughts and crosses in that the combinations of moves that define a complete game are near infinite. Humans playing chess look ahead at the possible moves, alternating between their own and their opponent's and searching for advantageous and potentially dangerous situations. This may be easy enough for a few moves but becomes increasingly difficult, as the number of possible moves expands exponentially. For years, many chess-playing computer programs have relied on their processing power to search a number of moves ahead more effectively than human opponents. Success in this is scarcely AI – more a matter of powerful processing. However, human chess playing can bring other, less mechanistic, skills to bear, such as: setting traps; sacrificing pieces and reaching a disadvantageous position to lead to a more advantageous position; seeking dominant positions over a field of play; and recalling previously successful patterns of play – all aspects of intelligent chess-playing. Replication of these tactics, by a computer program that can search effectively and carry out pattern recognition, indicates the use of AI. Indeed, the best computer programs can now play at international masters' level.

5.6.3 Poker

This is a card game with deceptively simple rules. Minimal skill is required to follow the rules, so that a small computer program could provide an adequate opponent. However, the pleasure and challenge of poker is in the process of misleading opponents, through bluff, double bluff, and beyond. Facial expression and body language play their part. For a computer program to be an effective poker-player would entail extraordinary powers of vision and perception. Image capture, repeated many times, would require scene interpretation and then gesture interpretation – well beyond the capabilities of current AI. The level of uncertainty in poker is extremely high.

5.6.4 *Robot football*

This has developed over a number of years, mainly in universities, and is run as a competitive 'sport'. It is, however, a serious area for research in AI as it poses a number of challenges related to uncertainty, as indeed does football itself. Robot footballers act autonomously: they are not guided 'telechirs' of the 'Robot Wars' variety. Once in play, their behaviour must be programmed: to locate the ball, move towards it if appropriate, pass the ball to another team-member, or kick it into the goal. AI is readily apparent in the required vision system, which must be able to identify the moving objects of ball, own and opposition team-members, and the stationary objects of goals and side walls. The rules of behaviour, which determine which mobile robot should chase the ball, in which direction it should move, and where it should aim the ball, are extremely complex, and certainly require the techniques of AI. Although the pitch and ball are of standard dimensions and surface finish, there is a high degree of uncertainty in the route a ball may take, its speed, and its deceleration. Positioning of the opposition robots is also uncertain and constantly changing. As football roboteers steadily improve the performance of robot footballers in their ability to accurately pass the ball and shoot for goal, they will increasingly come up against further uncertainty, namely the tactics employed by the opposition. Robot footballers that gain the ability to learn from the opposition's behaviour will show a high degree of AI.

5.6.5 *Industrial processes*

The degree of uncertainty may vary considerably in industrial processes. Many commercial systems can use information obtained on-line to control processes with techniques already described. An extension is to use both the direct information obtained from sensors and transducers together with derived information such as the differing time intervals between these events, as previously described in Chapter 2. With many processes, the time interval between events can give very important indications of malfunction, whether of a permanent or intermittent nature. This approach to understanding uncertainty in industrial processes, allied to intelligence (definable in many ways, but here related to the ability to handle data and incomplete information), is a powerful approach to being able to plan control actions.

5.7 Techniques of AI and how they are applied

There are many human capabilities that have their analogues in AI. In some cases AI can easily outperform humans, for example in rapid searching. In other cases, AI is no match for human skills, as for example in being able to interpret an unstructured visual scene. The human abilities relevant, as analogues, for intelligent machines include searching, pattern-matching, learning, decision making, dealing with approximations, vision, recognition and identification, and evolving solutions. These can be provided by AI, to a greater or lesser extent, in the following:

- Searching algorithms (including simulated annealing)
- Neural networks for pattern-matching, recognition, and learning
- Rule-based systems for decision-making
- Fuzzy logic for control purposes
- Image capture and interpretation for machine vision
- Genetic algorithms to improve pattern matching and searching processes.

What follows are a few examples of how the facilities might be applied.

5.7.1 Vision system for cheque reading

Humans read cheques to extract information such as the amount to be paid, but the process is slow, and prone to error and misinterpretation of bad writing. A faster and more accurate procedure involves the use of electronic image capture, pattern matching to identify features of the writing, and a neural network for recognition of numbers and letters.

5.7.2 Fuzzy control for central heating control

A traditional control system might work on the basis of simple rules which switch the central heating on and off according to ambient temperature. Fuzzy control might make the system more effective by being able to handle more variable concepts such as setting the heat control very high but for a short period, according to the time of day, personal preference, and whether the house is to be occupied for long.

5.7.3 Automatic lawn mower

Such a device, and there are a number currently available, needs to operate effectively and safely. Ideally it would be able to survey any shaped, continuous lawn and cut it without missing any patches. In practice, most shapes of lawn need to be indicated by a boundary wire, which is sensed by the device. There still remains the task of how to ensure that the whole area is mown in a methodical manner – not a trivial task for AI. The mower needs to be able to recognize obstacles which might cause damage, such as large stones, or which might be harmed, such as a domestic pet. Obstacles are a source of uncertainty and such mowers need to make decisions about how to avoid them.

5.7.4 *Intelligent leg-joint prosthesis*

Traditional artificial limbs can be tiring to wear and a poor compromise for the various ways a leg can be used, cycling and walking being very different, for example. A vision system can be used to identify ground position and whether it is flat or sloping. A neural learning system can identify the different types of movement and adapt microprocessor control of artificial joints.

5.8 Distributed intelligence using autonomous intelligent agents and the concept of emergent behaviour

5.8.1 *Agents*

The term 'agent' is widely used to describe a person or organization engaged to carry out a task for someone. An estate agent is entrusted with selling a house. A forwarding agent contracts to deliver a parcel. In both cases, the agent is instructed what to do but not how to do it. An agent may need to involve other agents, but that is not the concern of whoever commissions the task. In the context of the Internet, we are familiar with 'browsers' of the World Wide Web (WWW) – software agents, which locate information on request. It is a simple matter to request information about a book, an event, a company, a person, or whatever. The subsequent searching is a complex and sophisticated process, about which few users have, or need, any understanding. The 'web' browser is an autonomous agent that has the capability to understand the request, find the required information, and present it to the user in an appropriate form. Note, incidentally, that although a browser is a piece of software, it is difficult to explain its action using anything other than human-centred vocabulary, such as capability, understand, and present.

5.8.2 *Autonomous intelligent agents*

The concept of software agents has been around for a long time, certainly more than 15 years. Their real-life application is still in the early stages and is set to increase for many years. Web browsers are familiar examples of useful agents in that they are both 'intelligent' in being able to handle a wide range of tasks, and 'autonomous' in that they act with minimal instruction in carrying out complex tasks. As members of the class known as autonomous intelligent agents (AIAs), they are notably complex and rely on the use of very large databases. Many AIAs are simpler and may indeed comprise only a few lines of code and a few bytes of data, but are designed to operate within some degree of uncertainty.

Why is the concept becoming of greater significance and why is it of concern in this context? Over many years, computerized control systems have become more powerful, covering a wider scope of application and drawing upon ever-larger quantities of data. Examples include manufacturing resource planning systems (MRPII) which were developed over many years, particularly during the 1980s and 1990s. As large 'monolithic' systems, they became increasingly difficult and expensive to install and maintain, yet they were required to be more flexible in operation. While much was accomplished and accommodated by increasing computer processing and storage capabilities, it became increasingly clear that alternative approaches are required. The quest during the 1990s for distributed processing of information, complemented by rapidly developing communications networks, has contributed much to the achievement of flexible operation. Large, monolithic programs that once resided on powerful mainframe computers were redeveloped in fragmented form and distributed among networks of servers. Furthermore, the move towards distributed computing can be seen as accompanying and reflecting human control of systems, as they move from autocratic, centralized control to distributed control, in which decisions are made at the location where they are relevant. This leads to the idea that specialized, autonomous, intelligent agents can be used for localized decision-making, while operating within the wider context of a networked system. Extending the analogy even further, one can argue that it is easier and quicker to make rational decisions at the location where the situation can be best understood and the resulting outcome most effective.

5.8.3 Communities of agents and emergent properties

While this approach can be seen to solve some of the problems of multi-decision-making in an environment of constant change and uncertainty, it does beg other questions. How can one ensure that a collection of AIAs can work in harmony, helping to achieve the aims of the complete system? How can one avoid decision-making in one location that could affect other locations adversely? How can an effective community of AIAs be assembled and tested? How can it later evolve in order to accommodate changing requirements and external environment? There is no easy or automatic answer, but it is worth continuing the analogy with the control of human control systems. Consider the following principles for the behaviour of agents, both human and software. Agents must be able to:

1. Understand and act within the aims of the organization

2. Follow a number of rules, which define their behaviour and contribute towards achieving desired outcomes
3. Communicate with other agents to pass and request information
4. Negotiate effectively with other agents whenever a situation is of joint concern.

Additionally it would be useful if agents could modify their behaviour in the light of experience and changing conditions. This is the process of 'learning'. Human organizations rely on their agents (members or employees) being able to learn, in order to survive and prosper. Increasingly, AIAs are being designed to 'learn'.

In a human organization, if the criteria listed above are achieved then successful operation becomes more likely: a factory produces good products when required; a hospital treats patients efficiently; a football team wins matches. The concept of 'emergent property' (or externally observable behaviour) of the organization is a recognition that individual components or agents in the organization, which are specialized and limited in the tasks they carry out, nevertheless produce a result which is often surprisingly impressive. 'Impressive' usually implies a desired outcome, but the outcome could be impressively bad.

An example from nature is the termite 'palace'. This elaborate construction of tunnels is not the product of a termite head architect, chief engineer, and project manager, but the result of many thousands of termites following simple procedures in moving sand and helping each other. The termites are equivalent to simple AIAs, but the emergent property of their interaction is the incredibly complex termite palace.

5.8.4 The concept of proliferating simple autonomous agents to propagate emergent behaviour

While much has been written about the concept of assembling large groups of co-operating agents, referred to as 'swarms', to date little commercial success has been achieved. Two potential areas where developments are underway are in the concept of the Intelligent Geometry Compressor (referred to in Section 2.5.1) and in manufacturing supply chains.

In the intelligent geometry compressor, intelligent agents would be used to act on behalf of individual blades in rows of stator blades. They would obtain data from pressure (and other) sensors that relate to aerodynamic conditions, and pass messages to other agents about the changing conditions and receive messages from them. The result would be local decisions on how to adjust individual stator blades' angles in order to make them able to cope with local aerodynamic instabilities, thus making them more efficient in compressing the surrounding gas – the desired emergent behaviour.

In the case of desired emergent behaviour for manufacturing supply chains, what is required is the ability to satisfy customers' requirements in a situation of considerable uncertainty about supply and demand. The ever-changing state of customer demands has an effect on production along all the complex supply chains of materials, processes, and the components that contribute to the final product. This is achieved through negotiations along the many stages of supply chains, often driven by compromises such as a lower price for different car models. The proposal is that negotiation among intelligent agents communicating within a unified system over the complete supply-chain structure could provide faster and more effective control over what is produced. The potential for faster decision-making, reduction of waste, and closer attention to customer requirements would provide significant commercial advantage.

Software to support both these applications is available. Where the task now lies is in being able to implement and tune such systems so that they continually converge towards the desired emergent behaviour and not diverge towards malign behaviour. This is becoming a major issue in the use of large numbers of intelligent agents within complex human and mechanized systems.

5.9 Current technologies and their significance

5.9.1 *Neural networks*

Neural networks originated in the 1940s from concepts of neural activity in the brain. After many years of neglect, the ideas resurfaced and since the 1970s have been researched and developed as an important contribution to AI. Many applications are now being developed or are in use. The term 'network' is slightly misleading, since it is not related to a communications network but to the connections between the processing units which make up a piece of software known as a neural network.

Neural networks are used to *match patterns* and are thereby able to recognize or classify. They can carry out identification tasks by examining a range of features, and are useful where identification cannot be defined by simple rules. Consider a vision system required to recognize hand-written characters, such as the letters and numbers of a postal code on an addressed envelope. Identification could be rule-based if characters were printed in a uniform manner, but with handwriting, recognition has to take into account the careless shaping of features and a range of possible distortions. Human judgement is needed to distinguish, for example, between badly formed figures 2 and 7, or letters A and H. Analogously, neural networks can be used to make similar judgements.

Neural networks are in no way magical, but they are unusual in that their performance changes with time. Through the process of *training* they are said to *learn*. Training involves the presentation of inputs for which identification is given. The many and varied complex inner workings of neural networks, which are not described here, involve the automated adjustment of weightings applied to signals among the elements, or *processors*. As training proceeds, the intention is that the neural network should become better and better at carrying out accurate identification. In practice, varying degrees of success are achieved and in some cases recognition may be unacceptably poor. Trade-offs are often necessary between the training effort required and the degree of success acceptable for the application. As with human perception, neural networks are not infallible.

5.9.2 Image capture and interpretation for machine vision

Neural networks are associated in many applications with vision systems, in which a digitized image is captured and then analysed by *machine vision* software. This is a growing and significant area for development because it can potentially mimic many human activities that involve vision and the interpretation of what is viewed. However, constructing effective vision systems is fraught with technical problems.

The image captured of a two- or three-dimensional scene is represented as a simplified two-dimensional, digitized image, in the form a matrix of greyscale (or colour) readings for specified areas within the field of vision. With smaller areas, greater detail can be captured. An image can be output to a screen of pixels and viewed by humans, but any automated vision system must be capable of interpreting the matrix of readings. Human vision is capable of recognizing basic features such as straight lines, curves, circles, flat surfaces, rough surfaces, and corners. Automated recognition of such features requires complex, sophisticated algorithms to do the same task, and be sensitive to changes in the image environment, such as illumination, distance from the feature, and its orientation. In a scene of any complexity, recognition needs to go well beyond identification of features. Straight lines, curves, and surfaces can be combined to represent objects such as cubes and spheres, easily recognized by humans but a far from trivial task for machine vision to identify. More naturalistic images, such as components of a human face, similarly present difficulties in recognition. Machine vision is highly desirable for many applications but difficult to achieve.

5.9.3 Searching techniques

There are many software searching techniques in use, mirrored in physical processes and a number of activities encountered in everyday life.

Appropriate software has to be selected, just as we have to make tactical decisions about how to search for an object or a solution and how far it is sensible to pursue the best possible solution. The concept of a *search space* expresses the totality of possibilities within which a solution or solutions may or may not be found. There is a broad distinction between: exhaustive or *deterministic* methods, in which a *search space* is explored methodically; *heuristic* methods, which attempt short cuts using knowledge of the type of data stored; and *probabilistic* methods, which include elements of chance.

5.9.4 Deterministic methods

Brute force

This implies either searching every possibility in order to select the best available, or searching until the first satisfactory solution has been found. If it is known that a good solution does exist then brute-force searching may be appropriate, but the method may take too long or require excessive computing resources. Searching may be linear – working from the beginning of a continuous stream of data – but data may be arranged in other ways.

Tree searching

Data may be organized in the form of an inverted tree (or hierarchy). This may seem artificial, but it is not uncommon in the everyday world when representing situations in which a number of alternatives can each lead to further alternatives, and so on. Chess players explore possible moves, each of which lead to further moves, continuing the exploration to whatever level (or depth) the player can mentally perceive. A mechanism such as an articulated, robotic arm may have a range of initial positions based on movement of its first articulated limb. These positions allow a range of positions for a second limb and so on for all the limbs. Finally, the range of end positions depends on all the preceding limb positions, with the totality of positions being expressible as a hierarchy of related positions. Searching a hierarchy may progress *breadth first*, looking at all the first-level possibilities, before looking at all the second-level possibilities, continuing to the third, and so on. If there are many alternatives at each level, the search can quickly become exponentially large. Pursuing potential solutions by progressing through the levels – *depth first* – is a possible strategy if there is sufficient knowledge to allow such a directed search. In practice, hierarchic searches are often a combination of breadth and depth searches.

Heuristic searching

This approach is applicable when knowledge about a situation or system allows one to make short cuts in searching. Just as in everyday life, when locating a lost item is made easier by looking in specific places where it might be rather than systematically looking everywhere in the house, so heuristic searching involves the use of best guesses and 'rules of thumb'. The result may be speedy success but, as with all rules of thumb, there is no guarantee of success. Consider a vision system, which surveys a digital image in order to identify the presence of certain solid objects. Using a heuristic approach, objects such as cubes or spheres might be identified by searching for features such as straight edges and curves (by no means simple to do), thus achieving an acceptable success rate. However, unusual lighting conditions, object size, or object orientation might invalidate the heuristics and result in identification failures.

5.9.5 Probabilistic methods

Hill climbing

There are numerous variants of hill climbing, based on the concept of a landscape of hills and valleys, in which the tops of hills (or the bottoms of valleys) represent the positions of optimal solutions. Reaching these solutions is the process of recognizing the slope at any point and moving up or down the slope according to specified rules until a satisfactory solution is found. It is a matter of convention whether one ascends or descends to optimal solutions. Basic hill-climbing methods make the assumption that a solution can be found within the locality where the search is taking place. A local maximum (hill top) or minimum (valley bottom) may represent only a local optimal solution, whereas better solutions exist in other parts of the solution landscape. If there is a requirement for searching more widely, then other methods, such as genetic algorithms and simulated annealing, are required.

Genetic algorithms

These introduce a probabilistic element, which is analogous to the biological concept of genetic changes in successive generations, leading to survival of the fittest. Just as, in nature, pairs of parent chromosomes contribute to the chromosomes of their offspring, so variants of searching algorithms can be combined in order to improve their performance. An element of chance, analogous to the process of mutation in nature, is introduced through the use of random numbers.

Simulated annealing

Annealing is the metallurgical process in which metals such as steel are hardened by heating them to a high temperature and then slowly cooling them. The settling of metallic crystals into their lowest energy state has an analogy in the searching technique known as simulated annealing. Searching, which would otherwise take place within a local minimum (finding only a local, optimal solution), may be extended to other localities, opening up the possibility of finding a minimum which represents a better solution. A probabilistic element is introduced by continuously searching for solutions but gradually increasing the probability that a solution is to be considered acceptable. At the beginning of the search, analogous to the high-temperature state, potential solutions are unlikely to be accepted. As the 'temperature' decreases, the search continues and is increasingly associated with a particular minimum.

5.10 Rule-based systems for diagnosis and decision-making

For many years, rule-based techniques have been associated with *knowledge-based* or *expert* systems, which were intended to mimic the human ability to draw conclusions from limited and inexact information. In some situations, both in the everyday world and in engineering design, conclusions may be based on hard data and well-defined logical processes. However, in many cases, human judgement is called upon, and decisions are based on, heuristics or rules of thumb. While there is no guarantee of successful outcomes in all cases, the more and better the rules developed, the greater the likelihood of success. Rule-based systems are used to diagnose a range of possible conditions or draw conclusions about actions to be taken, by reference to a collection of rules which are applied to the data or information available. In most cases the rules could be invoked, or *fired*, in any order. In practice, the goal of the system determines the structuring of the rule base and the sequence in which rules are fired. It is expected that as use of the system produces information about its rate of success, the rules can be tuned, with the intention of improving performance.

5.11 Fuzzy logic for control purposes

Fuzzy logic is an attempt to handle the inexact nature of much data, and can be used in the process of *fuzzy control*. In many situations, where the parameters of a physical system are well understood, traditional control methods are appropriate. However, there are many situations where human control is less exact but is nevertheless successful in practice. These situations are characterized by inexact (even vague) expressions, such as 'fairly heavy', 'quite light', 'longish', and 'a bit' (as in 'turn it up a bit'), etc. Fuzzy logic depends on the arbitrary definition of such terms as 'heavy', 'light',' long', and 'a bit', in whatever way is deemed appropriate and of practical use. A simple example might be where a certain length is defined as 1.0 long and a smaller length as 0.7 long (that is, still long but not *as* long). Long and short may even overlap if, for example, 1.0 long is defined as the same as 0.9 short; that is, not very long is equivalent to not very short.

In fuzzy control, physical variables are first defined in this fuzzy manner, and are then used in a form of mathematical logic, which enables an exact (or *crisp*) value to be derived from them. For example, a central heating unit under fuzzy control might need to decide how far to turn up the heating control under the instruction:

If it's very cold outside, fairly cold inside, and I am going out soon (within the hour), turn up the heating a bit for ten minutes only.

Fuzzy logic systems can be applied to many situations, but their success is very dependent on the empirical process of designing and tuning them. As you would expect, there are no hard and fast rules about how fuzziness is defined in all situations.

5.12 How do these concepts fit into systems engineering?

The techniques of embedded artificial intelligence are required to fit into the principles of systems engineering. Important questions to ask are therefore:

- Will a traditional 'off-the-shelf' control system satisfy requirements?
- If not, what is the nature of the uncertainty that the system will encounter?
- What useful information can be sensed from the internal and external environment in order to reduce uncertainty?
- How can this information be used?
- What processing is involved with the information collected – searching, pattern matching, recognizing, classifying, interpreting visual images?
- How can the processed information be used to control actuation and the resulting behaviour of the system?

These are all questions that need to be asked at each stage of design.

5.13 Reference

(1) **Bissell, C. C.** (1994) *Control Engineering,* Stanley Thornes Pub. Limited.

Chapter 6

The Challenge of Idea Sourcing

This chapter is based on the keynote address given by Professor Chris Pearce FREng, FIMechE, FIEE at the Royal Academy of Engineering Visiting Professors Conference in 2002. The chapter outlines very succinctly the need for networking in multi-disciplinary design.

6.1 Competing in a knowledge-based world – the challenge of idea sourcing

Getting the right product to market at the right time lies at the heart of an organization's ability to compete and survive in today's environment. The quality of the insight determining what that product should be remains one of the crucial factors for organizations wishing to be successful. Knowledge of the market place informs the decision-making process and enables the company to reach appropriate conclusions. Competitive advantage has traditionally turned on the relationship of the company with its customer, and the perception the customer has of the company, relative to its competitor. There is a danger today that what the company does not know of technologies and methods available in other sectors represents a real threat to its product strategy. New entrants or competitors introducing new methods from elsewhere are also a danger. At the same time, there may be a potential competitive advantage to be exploited if methods can be successfully transferred to the company itself. We need to consider the potential of new technology and its transfer as a key corporate activity.

The development of a technology strategy should therefore contain a new dimension. The analysis of the company's strengths, core competencies, and why it is successful in the market place must be complemented by a new interpretative stage, where the trends in factors such as legislation and technology are considered before determining how technology is to be managed within the company. This key stage changes the new product introduction process and needs a new form of vision.

The challenge is to develop the ability to find and link appropriate technologies key to the company's products. This implies an understanding of which technologies to use and where they are to be sourced, combined with the understanding and vision to track developing technologies, introducing them at the appropriate time and degree of maturity. A technology manager is a must.

These are very important strategic questions for companies to answer and analogies can be helpful in thinking about how a company relates to its technology. One such model, the 'bonsai tree' concept of a company and its product knowledge base, has been suggested in the past (Fig. 6.1). This model proposes that the roots of the tree are founded in generic technologies, the trunk represents the core technologies of the business, and the fruits the products of an organization. The problem with this model is that the roots of the bonsai are constrained by the pot and by deliberate pruning, resulting in a beautiful, but stunted, organism. This is the model of an organization finding its technology only in known and comfortable, traditional sources. In order to break away from this closed view of a business, it is more appropriate to consider the Californian sequoia tree as a model for the sourcing of technology and the technology transfer process.

The natural history of the sequoia is instructive and worth understanding, to put the proposed representation of the business world into context. Each tree in the sequoia grove is an individual, but whenever the

Fig. 6.1 Bonsai tree

roots come into contact with the roots of another, fusion occurs and the trees share the extended root system. Adjacent trees grow together where they touch, resulting in two separate, individual canopies with a single trunk. During the wet season, each root system will store large volumes of water, and this is released progressively during dry periods. What is interesting is that the interconnecting root system means that a tree in a drier area can in effect draw on a communal resource in times of need.

This observation leads to the view that perhaps the 'sequoia model' is a better way to describe the relationships between companies and industries at the technological level (Fig. 6.2). A company can grow its products in relative isolation, developing its generic skills, its core knowledge, and its products. This model is the 'bonsai', a small-world view. The development of technology outside a sector can have significant impact when transferred in, and the development of complex supplier chains suggests that today's environment requires the large-world perspective of the sequoia model.

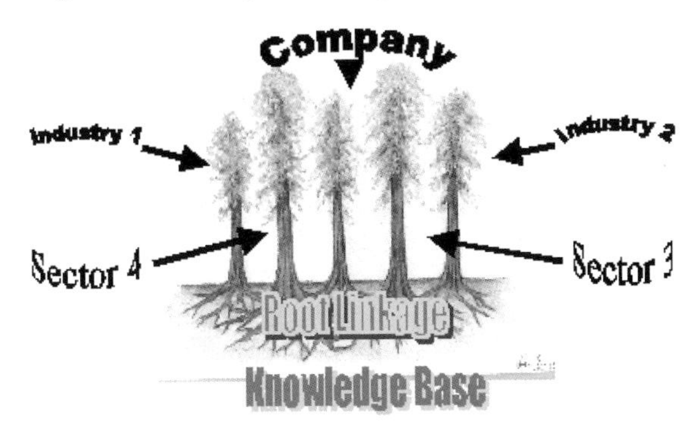

Fig. 6.2 Sequoia tree

In this model, the role of the individual responsible for the strategic development of technology within an entity is to determine where and in which industries or supply bases the key technologies needed for future success can be found. The role is to act as a means to 'touch' the roots of the next-door tree and make the connection. In this way, the roots of the organization are in the wider-world knowledge network and can provide the support benefits needed for today's competitive environment. The sequoia has no enemies other than man and heavy snowfall, and can live almost indefinitely. As the objective of strategic management of technology is the long-term survival of the organization, the model is a good one to consider, given that, increasingly, organizations cannot develop all their own technology and must necessarily outsource skills and technology.

Working closely with a major INBIS client, BNFL, a niche requirement for very-high-power manipulators was identified. In responding to this requirement, INBIS developed the underlying technology and worked with the client to develop a machine to meet the need. The resulting robot is a very highly loaded, seven-degrees-of-freedom, remotely operated machine used for materials-handling in the nuclear decommissioning sector. The challenge was to develop a machine capable of being deployed through a 270 mm diameter access port, capable of manipulating a 120 kg payload at 2.3 m arm extension, and modular in design to allow easy reconfiguration – up to a 4.6 m extension. Fundamental to the requirement was the ability to design hydraulic rotary actuators subject to enormous loading, with clean lines for decontamination reasons, the ability to be deployed and recovered through small access ports, and with the reliability necessary for application in a nuclear environment.

It is interesting to consider this challenge in the light of the sequoia tree model proposed earlier. The need was identified through the customer link. The analogy is that a water shortage was detected in the customer tree. The means to partially satisfy the need was available within INBIS Nuclear Division (namely, its experience of manipulator design within a nuclear environment), but the total means was not. There was insufficient knowledge of exotic materials and advanced analytical techniques within the nuclear robot design team to prove whether the design brief could be achieved. In this context, the role of the technology manager was to recognize the nature of the shortfall in knowledge and to establish access to the relevant technology.

The 'missing technologies' in this case were to be found in several key areas, in particular in the aerospace sector. The technology manager had a strong aerospace background, as well as exposure to ceramic bearing technology, and was able to make the necessary connections. It was the establishment of these connections within other technology sectors that enabled the requisite innovation to be realized. This deliberate decision on technology sourcing resulted in an actuator concept using aerospace and waste-pumping materials (titanium, maraging steels, and ceramics) and aerospace analytical techniques (FE modelling for stress, strain, and life) to supplement the nuclear robotic design knowledge. This combination of technologies, commonplace in their own fields but not applied together before in the recipient sector, achieved the necessary breakthrough.

An interesting technology sub-root system has also been identified in this model. A key technology in manipulator design is the development of the end effectors, particularly the grippers. These end effectors have a heavy duty, requiring parallel motion of the jaws, while maintaining sufficient force across the full range of jaw openings to ensure that the load is always safely carried with no danger of falling out in transit.

Over seven years ago, INBIS identified an interesting technology being developed in the field of constraints modelling of high-speed machines. Working in partnership with the University of Bath, INBIS has developed a software suite from the results of this research known as INca® (INBIS Constraints Analyser), which allows designers to explore the constraints within a design process and optimize solutions in complex multivariate situations. Gripper design has been a core robotic skill at INBIS for over 20 years, and initially the design followed normal practice. In this approach, the designer concentrates on three key variables and aims at achieving a design solution at a minimum of three positions during the jaw-opening cycle. This is typical of complex constraints in design problems, where designers may be faced with many variables but concentrate on those which experience tells them can be manipulated most easily to move towards a solution. Many possible solutions are inevitably not evaluated due to the time constraints.

The use of the constraints analysis approach showed that the designer was really working with 11 major variables (some of which were geometric, others economic), and that the existing design could be further improved and sensitivity to manufacturing tolerances reduced. A new solution was developed using INca® technology that resulted in changes to a number of variables. The changes were, on the face of it, small. This is not surprising given the pre-existing experience of such designs. The new design had joint-pivot positions changed by fractions of a millimetre, crank angles by 1–3 degrees and other minor modifications (Fig. 6.3). The changes, however,

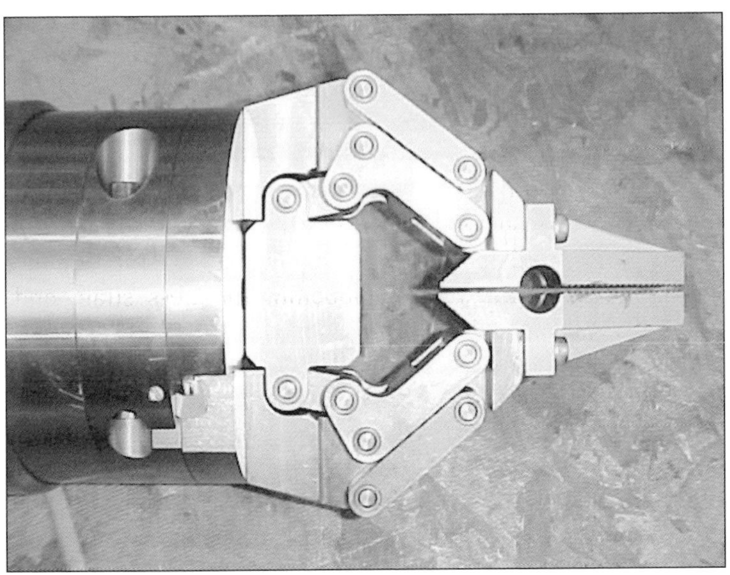

Fig. 6.3 Parallel jaw gripper

resulted in a radical improvement in joint performance. The resulting gripper has: greatly improved efficiency, being both lighter and with very good performance under load; excellent manufacturability, with considerably reduced demands for extremely accurate component tolerances; and has been proven in operation to be extremely reliable. In this technology transfer, the role played by a University of Bath/INBIS Teaching Company Scheme was excellent, and proved a very effective method to apply new thinking to traditional processes.

The crucial role in enabling the sequoia model is that of the technology manager, who must actively look for opportunities to create the links at the roots of the organization. Most innovations contain a breakthrough requiring knowledge from different industrial sector or fields of science. How do we train people to look to make such links?

6.2 Mechanisms that can assist technology awareness and transfer

There is considerable historic evidence that technology transfer between sectors has happened more by accident than design. The challenge for today's managers is to embody into their planning process an awareness of the possibilities available across the breadth of technology, and to be open-minded enough to recognize that not all wisdom is resident in their own organizations. The first key step is top-level commitment to develop awareness and act upon the findings. Without it, progress will not be made.

A number of methods can be used to develop awareness. One successful strategy is to give key individuals responsibility for watching for developments in new sectors. This 'technology watcher' activity can inform the organization and enhance awareness. Recruiting from organizations in different sectors can be helpful, with new knowledge being imported with the staff. Linking with academic sources can be very effective, provided that the relationship is handled well and the industrial partner recognizes the differing drivers that exist in academia. Partnering with others can bring the complementary knowledge to move forward.

Smaller organizations may need to rely more heavily on external sources of help. In this regard public bodies can play a part in developing, for example, subsidized consultancy services, open information support services, business-link-enabling mechanisms, and so on, all of which can encourage smaller organizations to look outside their comfortable and familiar technologies. Large private companies can have an influence in raising awareness of the issues by promoting the involvement of smaller companies in their supply chain.

Modern IT systems can also assist, both in capturing what an organization already knows and making it easily accessible for future re-use, and in accessing data sources. To this end, training in data search techniques can be beneficial and give quick results.

It is essential that multi-disciplinary product-realization teams operating across cross-sectoral boundaries are enabled, through the delivery from academia of well-trained young engineers with the confidence to look outside their narrow core disciplines into the wider pool of knowledge available today. Indeed, it is possible to argue that the emerging engineering cadre is best placed of all to react to the challenge of sequoia sourcing of technology. The information age has been with them right through their education, and their facility with data access and mining far exceeds that of most of their peers in industry. They should emerge with a natural advantage and predilection for technology transfer activities.

Young engineers need the experience of working in multi-disciplinary teams – a powerful formative experience. The importance of such experience has long been recognized, but its delivery within academic environments has been patchy. The existence of the Royal Academy of Engineering's Visiting Professors in Principles of Engineering Design Scheme has been a positive response to seek to correct this problem.

Successful models have been developed wherein student engineers work together as groups to deliver a design project, ranging from a complete aircraft to more modest-scale tasks. These are excellent in concept, but are limited in the value of the experience because they are not truly multi-disciplinary. This latter dimension is introduced by visiting speakers providing relevant input, but there are inevitable limits that arise because more often than not all active participants are drawn from a single course. To best emulate the real world, such projects should be at least cross-faculty, bringing together student engineers from differing disciplines. There should ideally be a way to involve students from business and management studies as participating team members. The major blockage to this happening is usually the timetable and inter-faculty suspicion. Surely this cannot be allowed to be a significant reason why we do not strive to improve the way we prepare the next generation of technological entrepreneur? If we can organize the delivery of complex technology, surely it is not impossible to organize the delivery of more focused education for our future engineering practitioners?

We need to develop good-quality case-study material in this area, as it is only through example that we will be able to convince both industry and academia of the benefits that can be delivered through innovative technology transfer.

Today's students are being educated in an information-rich age. A key skill is the ability to data mine, and this will assist them to contribute effectively to their new employers' search for competitive products. They will, however, need to learn perseverance and persuasion in seeking to source new ideas from other sectors, as they will find resistance to change.

6.3 In conclusion

The capture, analysis and recycling of knowledge will become ever-more important, and will need to be managed as a critical organizational resource. This cannot be dealt with purely as an information technology issue, as we are dealing with human beings interacting with the systems that we will evolve. Therefore it is essential that we design our management philosophies to satisfy human needs as a key part of delivering organizational goals. People working in virtual teams will still have their social needs and they will still behave illogically from time to time, but it is they who will supply the innovative leap. Our efforts should aim to provide a better environment in which the designer and manufacturing engineer can explore ideas creating new, innovative products, recycling information, learning from the mistakes of the past, and making the key connections across sectors that will provide the new products of the future. The role of the technology manager will be increasingly essential to the health of an organization. It is the training and formation of such managers that must be addressed, and it is in this area that the role of the universities lies.

The future of a United Kingdom seeking to be successful in an information-rich world lies in solving these problems. The ability of the people in an organization to import new thinking and apply it to their product environment is the key differentiator for the future. The vision of the company's senior management in creating a supportive environment for creative people will be closely linked to its long-term prospects, but even more vital is the ability of universities to engender in their graduate output the spark of awareness of what might be achieved through appropriate technology transfer. The vision and skillset given to young engineers emerging from the academic establishments will, at a national level, represent the driving force enabling the next generation to influence the current one in a positive and dynamic way. Achieving such a change at local and national levels is a new type of culture-change process with far-reaching consequences for all.

A final thought on the sequoia model is instructive. The forest fire is an essential part of the regeneration of the forest and in particular creates the environment for new sequoias to grow. The sequoia is virtually immune to fire because of the thick, protective bark with high tannin levels its trunk carries. However, fire is an essential part of its reproductive process. New sequoia seedlings only germinate in the fertile tilth left after a forest fire, when they have no competition from the undergrowth swept away in the conflagration. Seeds can lie dormant for many years, awaiting the heat of a fire to initiate growth. When big enough they make their own root system connections. If we get our connections right, our own organizations will be better placed to withstand the vagaries, ill winds, and 'forest fires' of the market place, and to put on new growth in response (Fig. 6.4). It is therefore imperative that we develop organizations with the agility to move technologies around and make innovative connections. Those who succeed will be the 'sequoia' organizations of the future and will have true longevity in the market place.

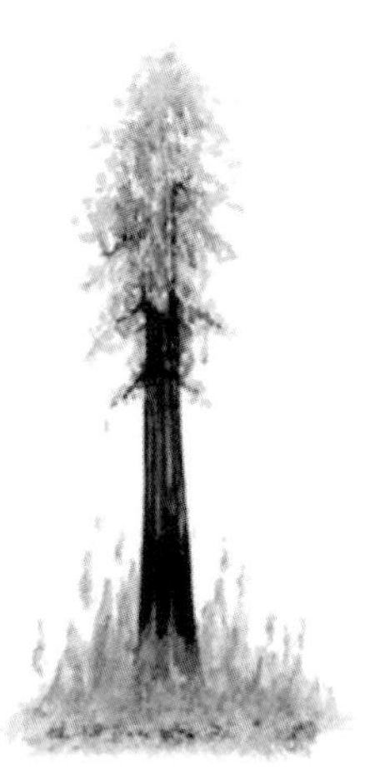

Fig. 6.4 Sequoia tree fire

Chapter 7

Case Studies

This chapter contains a number of case studies that are applicable to the subject of systems engineering. Not all the case studies have answers or outcomes, as they are meant to make the reader think, not only of possible solutions but also about the sort of problems that can or may be solved by a systems engineering approach. Some of the case studies merely illustrate how incorrect information can lead to totally inappropriate actions. These are included because at first reading they can seem unbelievable; however, they are all true. We must realize that negative information is also very useful!

7.1 The wrapping machine

Everyone is familiar with the way in which modern consumer products are wrapped and in some cases over-wrapped, that is, there is more than one layer of wrapping material. The machines that are used for these wrapping processes are many and varied, depending on the product, but in our case study we will look at the wrapping of a thin film of polyester around a 'hard pack' packet of cigarettes. This is a wrapped product that is familiar to us all, and the resulting wrapping is sufficiently rigid to exist as a sized envelope even when the cigarette carton has been removed, despite the flimsiness of the polyester material. The machines that carry out this over-wrapping operation have evolved over very many years, based on an original design where all the movements and wrapping operations required are obtained by purely mechanical mechanisms (see Figs 7.1–7.3). The basic concept of the design relies on the film being stationary, drawn across the mouth of a pocket, into which the product is pushed or 'plunged'. The action of pushing

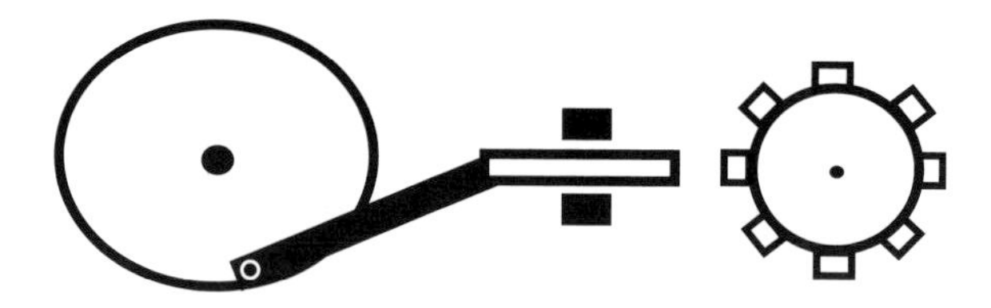

Fig. 7.1 Wrapper – index head principle

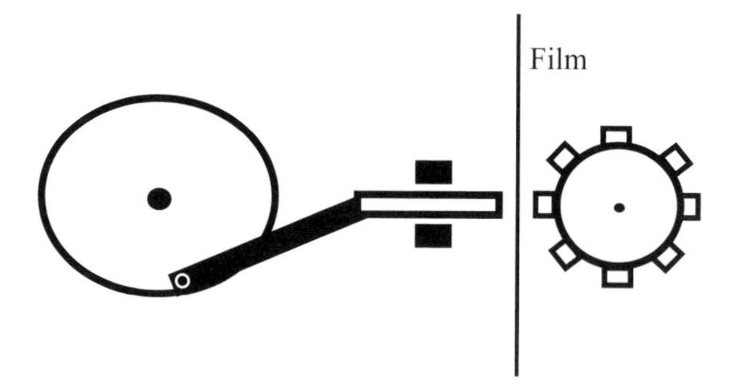

Fig. 7.2 Wrapper – index head with film

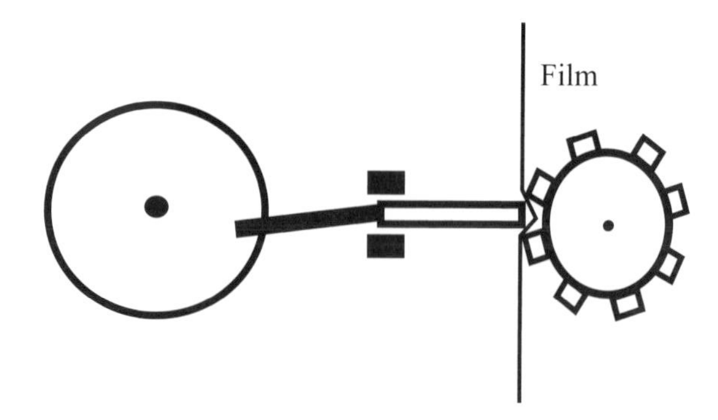

Fig. 7.3 Wrapper – film and pusher in pocket

the product (here a small box or carton) into the pocket, against the film, wraps the film around the product in a satisfactory manner. The neatness of the film wrap is most important from a quality viewpoint, and the design of the pocket and the plunger mechanism has evolved over many years to produce a high-quality fold (look at any cigarette packet). The important aspect of this operation is that the operation is carried out into a stationary receiving pocket, so the product has to be moved accurately into the correct position, brought to a standstill in this 'work station' and then moved on to the next work station. Here the film is heat-sealed by the use of heater strips, which are again stationary. The product is then moved to an inspection and ejection area, which may again be stationary.

The essence of this design approach is obviously that the wrapping and sealing operations require the product to be stationary in space, then accelerated and stopped at the next work station. The mechanisms required to achieve this, at the required accuracy, are complex and require high precision in manufacture. The operation is also inherently noisy. With the never-ending quest for higher operating speeds and improved quality, the complexity of these machines has increased to the point where a cigarette packet wrapping machine can weigh up to 9 tonnes and require a 5 kW motor to drive it (Fig. 7.4). This just to wrap a thin film around a small card-package! This type of machine also has a limiting speed of 450–475 p.p.m. (packs per minute). This is because the actual maximum speed is determined solely by the acceleration the product can withstand as it is moved from one work station to the next, and the maximum temperature the product can tolerate (remembering that the heat seal is time/temperature dependent). The design of such a machine is totally within the capability of 'mechanical' designers, since the size of the required driving motor is calculated from the mechanical drives, and a constant-speed electric motor is needed which can be easily chosen from commercial catalogues. It is also important to remember that this type of machine is a constant-speed device, and requires

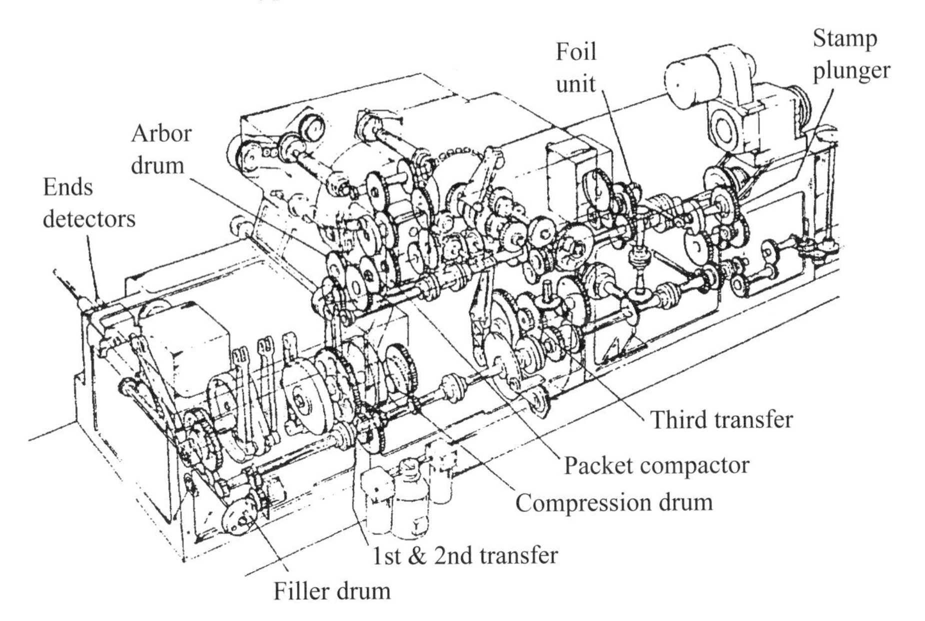

Fig. 7.4 Typical mechanical drives system

very extensive and time-consuming mechanical alterations to make a 'size change', that is to make the machine capable of accepting a different-sized product. If we analyse the development of this type of machine, we see that it has progressed along a purely mechanical design philosophy, with the basic principle outlined above developed to the maximum speed possible by refining the accuracies and profiles of the mechanical parts. The design is predicated around the drive transmission system rather than around the needs of the product, until a speed limit is reached based purely on the loads and temperatures put on the product itself. While 'electronic' controls have been added at a late stage, in the form of a simple PLC, this control does little but replicate existing mechanical systems, providing a certain degree of simplification but not improving the machine's capabilities to any great extent.

This type of wrapping machine illustrates perfectly the evolution of design through a mechanical engineering approach. Customer demands have been met by refining the accuracy of its component parts, necessitating more robust mechanisms to position them to the required accuracies (to minimize deflections), and leading to increases in weight and power requirements until we meet a physical impasse due to the characteristics of the product material. Symbolically, we can show this approach to the drive configuration in Fig. 7.5, which shows diagrammatically the drive to the two movements in the turret station being obtained from one drive motor, via mechanical gearboxes and actuators. Diagrammatically, the new total control and drive layout is as shown in Fig. 7.6, which should be self-explanatory.

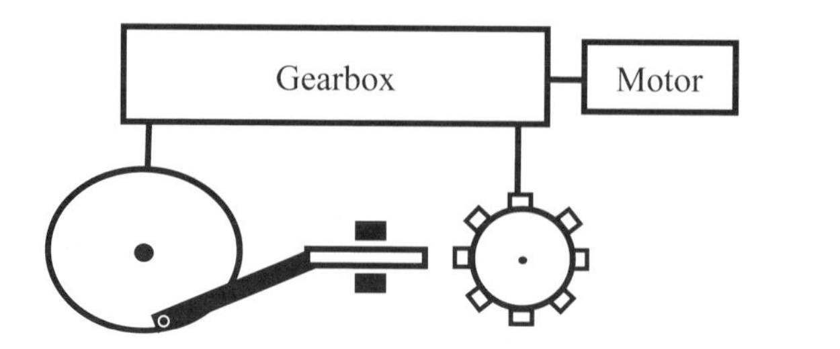

Fig. 7.5 Phase synchronization of drives

What do we do next, as the customer demands faster and faster product lines? The conventional answer is to duplicate the lines but the new technologies take us in a totally different direction. We can see that the existing approach has compromised the actual functions we need to perform

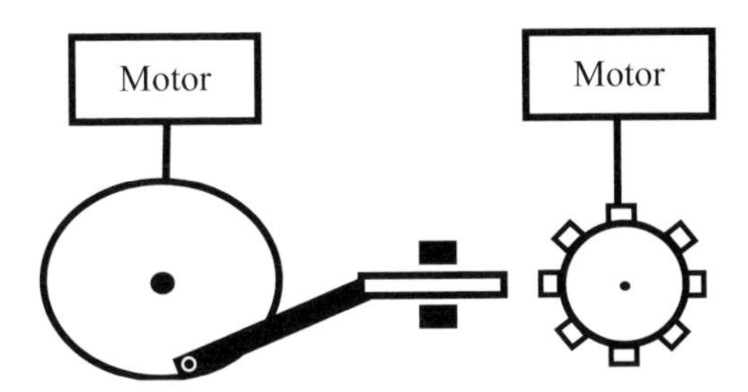

Fig. 7.6 Phase synchronization of drives – new approach

on the product itself, to achieve a satisfactory drive system for the machine. This has eventually and inevitably led to an absolute speed limit on the machine. This is not a criticism; it is a statement of where the machine design has led us from the initial design of 60 or 70 years ago, which has now been developed to its absolute limit. Perhaps Mechatronics has arrived just in time! To complete this case study, let us look at how a new machine was designed by taking a systems engineering approach.

In looking at the actions necessary to wrap a thin film of polypropylene around a cardboard packet, first we need to keep the handling of the basic materials to a minimum, so as not to damage them. We have seen that the existing machine is limited not by the mechanical design of the machine itself (although higher speeds require greater accuracy in construction, and there is an obvious final limit here as well), but by the accelerations and temperatures imposed on the product materials themselves. As a consequence, any new machine must remove or reduce these limitations by changing these parameters. At the same time, the existing design necessitates frequent changes in direction of the product that can adversely affect the essential customer requirement of product quality. Another requirement we need to address is the need to accommodate product size changes as quickly as possible, preferably without physical alterations to the machine itself and ideally during machine operation, known colloquially as 'on the fly'. As we have seen, size-changing with existing machines necessitates the physical changing of the parts within each work station, which requires the work of maintenance fitters and consequent 'down time', plus the provision of alternative-size parts to be kept in store and thus in inventory. This then is the brief; how do we tackle it?

The first consideration must be to be as 'gentle' with the product as possible, to enhance quality. Hence accelerations, temperatures, and changes of direction need to be kept to a minimum. The optimum here is to keep the product moving in a straight line, and this can be achieved with moving belts with speed variations determined by controlled motors. Individual motors can be used since any synchronization required will be achieved through software rather than rigid mechanical drives. The folding of the polypropylene wrapping can be carried out by appropriately shaped guides, between which the product is moved by the belts. Separate sets of belts incorporate heating elements, which produce the heat seal. By altering the length of belts we can choose the time the product is in contact with the heating belts, and thus the temperature for the heat seal can be reduced (remembering that the heat seal has a time/temperature characteristic).

The above is a brief description of an experimental machine that was built and run. The design is predicated on the use of software-controlled drive motors and the requirement to keep movements of the product to a minimum. The advantages of this approach over the standard design of over-wrapper are:

- Accelerations on the product are reduced to virtually zero.
- The time for the heat seal is increased so the temperature can be reduced.
- Size changes can be accommodated 'on the fly' via software, eliminating change-over time and the need to stock 'change parts'.
- Operating speed is increased from a limit of 475 p.p.m. to over 1200 p.p.m.
- Weight is reduced from over 9 tonnes to well under half a tonne, reducing floor loading.
- Power requirements are drastically reduced.
- Floor size 'foot print' is reduced.
- Capability of over-wrapping two or more packets in one wrap is possible (impossible with the existing design).

This is an impressive list. In this case study, the key to the change in technology and the resulting benefits is the use of the controlled drive motors and actuators. This approach to design of this type of machine is shown in the outline drawing of Fig. 7.7. In making this type of design change, there are a number of criteria that will have to be taken into consideration. The use of individual drives means that the inertia of the drive system will be significantly reduced. While this is advantageous in respect of power, weight, cost, and heat dissipation it does have an effect on the compliance of the various mechanical sub-systems in the design; the matching of inertias, compliance in couplings, etc. become much more important than before. Inevitably there is also a downside. Initial unfamiliarity of the design team with multi-disciplinary designs, unfamiliarity of the maintenance teams with

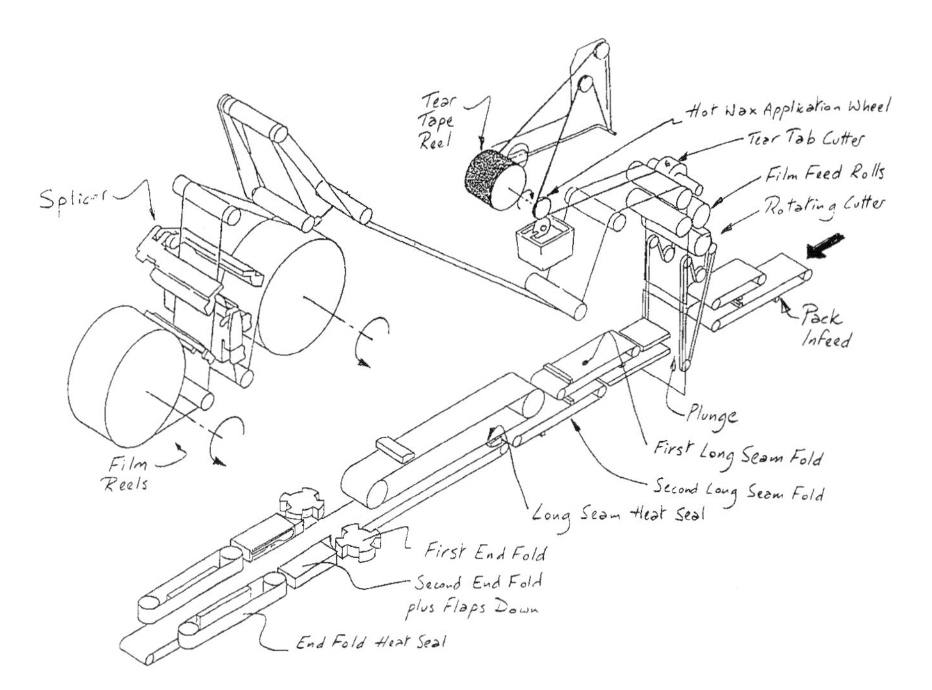

Fig. 7.7 High-speed wrapping machine (isometric sketch)

the new technology motors and software, retraining of operators, and so on all have an impact, but this should be of only a temporary nature.

7.2 The goods distribution system

Goods distribution systems are examples of a dynamic network, which have well-established precedents in the form of postal systems and public transport with trains, buses, and coaches, etc. that all run to well-established timetables. If the requirements are constant, timetabling can work well, but what are the real requirements in the distribution of goods and services?

Timetabling basically concentrates on organizing an arrangement of regular movements around a network. Whether the particular movement is required at a particular point in time, or even if the particular movement is over-subscribed, is not a feature of the schedule. It is assumed that the required movements will match the scheduled movements. Good examples here are train and bus services. The buses and the trains run to their timetables whether they are full or empty, whether there are no passengers needing the service or the number of passengers is twice the available capacity. Even with very complex networks, such as the early rail system in the UK (as distinct from the existing, abbreviated network), this type of network can be scheduled manually, although the use of a computer obviously should help. The early rail timetables are a beautiful example of what can be achieved by manual means, and the early copies of Bradshaw's National Timetables have become collector's items. This type of scheduling has become very common, but how can we match movement to demand?

In the air freight business, the transport costs are high, and the real need is to balance the available capacity and journey end-points with the demand. At the present time, while demand is high it is not really high enough to overcome the cost implications of running a timetable service. A much wider look at scheduling is needed, and looking at the whole from a systems engineering viewpoint we see that in this case we need to combine a number of different requirements (capacity, placement, time, and movements) in one network. Timetabling is too simplistic an approach, since it only tackles two requirements (time and movement). To cope with this amount of complexity, a multi-agent system (MAS) approach will have to be used. Although MAS development is still in progress, it offers the only real solution to problems which involve very many variables when operating in conditions of uncertainty arising from unpredictable events.

There are a number of other types of operations (such as the mail network) that operate on a timetable basis but which might well be organized more efficiently were some scheduling technique to become available. Similarly, if environmental aspects become even more important, the carriage of goods around the road networks (whether at a national or international level) needs scheduling if ultimate chaos due to further road congestion is to be avoided. This is a problem that is well recognized but which needs a systems approach (to bring in all requirements); however, there is not an answer for this at this point in time.

7.3 The local bus network

It is instructive to look at one example of timetabling that will be familiar to most readers. If we look at a local bus service in any town or city, we see a timetabling system being worked in a way that largely defeats the real requirements of a systems approach. The practices are so widespread, if not indeed universal, that we have come to accept them and the consequences of heightened road congestion. Bus services are intended to work to a timetable. As outlined above, this should mean that at a given point in time individual buses will be at known points on their journey. After all, if an intending passenger looks at the timetable it is to gauge how long it will be in time before the next bus passes. This is exactly how the railway operates, but a bus service assumes that if a bus arrives at a stop and there is no-one waiting to board and no-one wanting to alight, the bus can (and does) omit that stop. Further, in a unique example of parochial accountancy, the decision was taken many years ago to eliminate the conductor (or second operator) on the bus and to get the driver to take the fares and issue tickets. This immediately meant that the time at each stop is a variable (the antithesis of accurate timetabling) and, worse, the paying passengers are now expected to queue outside the vehicle, in perhaps inclement weather, while the driver issues tickets and gives change when required. What happens to the other road users? They queue up behind the stationary bus, causing further road congestion and pollution. In the meantime, the following bus is running ahead of schedule because it is missing out stops – because the first vehicle has picked up the waiting passengers. The time intervals between vehicles has now shortened and can (and does) lead to the well-known phenomenon of a longer wait before the first bus appears and then two or three arriving at once.

Since this type of operation can be observed first-hand on any day in our high streets, it is instructive to look at how a more efficient system could be run. From a systems point of view, the local bus company decision to go to one-man buses, rather than use separate driver and conductor, has obvious financial advantages to them in cutting costs. But in an overall systems concept it has increased costs (for everyone else), added to pollution, inconvenienced the fare-paying passengers and in particular has contributed greatly to the public's view that public transport is slow and unreliable. Perhaps if the transport authorities looked on the operation of a local bus service from an overall systems perspective, rather than at the bus operators' detailed view alone, matters might improve. Again, this study highlights how local financial requirements can influence systems judgements on a much larger scale, a theme explored later in this chapter.

7.4 Manufacturing systems and development requirements

In a manufacturing plant, particularly one centred on small-batch production, it is quite common to plan a number of alternative machine-loading itineraries to enable batches to be put into work on the machine cells available, rather than using the optimum process. With just-in-time (JIT) techniques, this dynamic scheduling is frequently done manually, of necessity, but systems engineering and MAS should enable on-line active scheduling to be achieved in the future. This, together with autonomous transport around the factory, should provide an optimum solution to machine-cell loading problems. However, there is one prevalent practice which can and does defeat the best systems approaches in manufacturing –

the need for the manufacture of small quantities of development parts. The following scenario has been witnessed twice in two of our premier and most up-to-date manufacturing companies.

In companies where the development of new products is carried out alongside in-house manufacture of production goods, there is frequent pressure to carry out both types of manufacture in the same plant. Initially it is seen that the needs of the development programmes can best be met by setting up a separate development manufacturing unit. Since the whole essence of this approach is speed of response, the manufacturing unit has to be equipped to cope with the predicted peak-load requirement capacity rather than the computed average-load requirement. Inevitably, this means that all the manufacturing capacity will not be used all the time and some machines and operators will be on 'waiting time'. Eventually it is decided to load production work into the development shop, to improve the machine-loading results and hence reduce manufacturing costs. Sooner or later, the machines now loaded with production work are needed urgently for a short-term development task. Should the production order be taken off the machine to enable the rush development task to be undertaken? Or should the development job (which is, after all, small-quantity short-time-scale work) be done elsewhere? Because of delivery promises made to production, the decision is frequently taken to off-load the work, either to a sub-contractor or (believe it or not) back to production! As this scenario develops over a protracted period, it is not unknown for the development shop to be loaded mainly with production manufacture and the manufacturing department to be constantly disrupted by urgent small-batch development work. The rationale behind having a development manufacturing capability is now lost and a decision is then taken (on financial grounds) to absorb the development capability into production.

At one plant that went through this process, the main production was run on a very effective MRPII scheduling system, with well-controlled loading of the machine cells. However, output was always behind the predicted schedule; the reason was eventually traced to the need to break down the production schedule every night to machine up to 600 development parts, needed for the next day. The decision was then taken to set up a new development manufacturing facility – full circle in 20 years!

The point of this study is the need to incorporate all requirements in a systems approach, and not take local and perhaps short-term steps to overcome basic and long-term problems. Not every systems approach will provide the most effective and lowest-cost solution to an overall problem when studies are made minutely at the sub-system level. In this case, although the cost of having unused manufacturing capacity in the development shop was high, for the overall system it was the correct and most cost-effective solution, as well as being the most efficient.

7.5 The missing audit trail

A company making very complex processing machinery was criticized by their customer for supplying machines of differing build standards. A check was made on the design specification for the particular machine type, and the standard appeared to be adequately specified. The assembly shop was asked to produce the final specification check-list for each individual machine, but was unable to do so. The assembly shop manager assured the technical director that the specification check had been carried out and that everything was in order, but he explained that the specifications and check-lists were kept on a PC. Their method of checking was to delete an item from the check-list when a physical check on a machine showed that a modification or part was incorporated. This was the easiest way on the PC. Obviously, this technique produced the result that at the end of the check there was no record of anything! Also, any additional modifications incorporated earlier than expected (which can happen, due to rescheduling) were not checked or identified. The easiest system was inadequate, to put it kindly.

7.6 Designer knows best

The design of the turbine blades in a gas turbine sets the performance limit of the whole engine, since the hotter the turbine can be made to operate, the higher the thermodynamic performance that can be achieved. The design is a difficult compromise of mechanical (strength) requirements, vibration reserve, and the aerodynamic and thermodynamic performance requirements. In this particular case, the requirements had been achieved by a complex design involving the use of high-strength (but difficult to machine) nickel–chrome alloys, with a multitude of small internal cooling passages that had to be drilled very accurately in the blade aerofoil.

The manufacturing department was having very great difficulty in making these blades; in fact every blade of the thousands made had some errors in it. In desperation, the production engineer concerned went to the design office and spoke to the chief designer, asking for easements in the design to assist with the manufacture. The response from the chief designer was 'I design blades, you make them. You go and do your job and don't worry me with your problems'.

This case illustrates a point in multi-discipline design that is even more important than before. That is, that everybody's actions influence some other aspect of the design. In the case illustrated, a culture had grown up that the design office were experts in every aspect of engineering and therefore their decisions and designs should not be questioned. This point has been discussed in earlier chapters on design evolution. Although this example is not new and concurrent engineering techniques have done much to change attitudes, with today's knowledge this type of attitude must be seen as unacceptable, but it is surprising how frequently it exists. Multi-disciplinary design *demands* teamwork.

7.7 Electric brakes

A recent development in the automobile world is the introduction of electric brakes. In this system, the normal hydraulic-actuated calliper is replaced by a calliper incorporating electrically driven actuators that provide the clamping load onto the brake disc. Control is through 'brake-by-wire' systems, which can be incorporated into the overall vehicle system to give an integrated braking system.

What are the advantages of electric braking systems? First and foremost, there is much easier integration into the overall vehicle control system. Secondly, there is reduced fitting time in assembly plant plus a cleaner system by the elimination of hydraulic fluids. Long-term reliability, however, is yet to be proven, with the well-known difficulties with electrical systems in motor vehicles. Corrosion in connectors, hardening of wiring, etc., are all areas which will have to be proved by vehicle use and are already of major concern in aerospace, where higher standards apply. These are important aspects of the whole-life approach to systems engineering, and aspects of this particular approach to braking that have still to be satisfactorily answered.

7.8 Agriculture – farming as a system

7.8.1 The automatic milking machine

A research project at Silsoe in the late 1980s aimed to produce an entirely automatic milking machine that would enable cows to be milked at any time when they so desired. The project was obviously multi-disciplined, incorporating sensors and robots and an intelligent control system. There was an early requirement to be able to recognize individual cows; the 'datum' for the cow's length was its mouth, as the attraction of the feeding trough was one inducement to get the cows to co-operate. A sensor was therefore attached to each cow, which transmitted the cow's identity to the system, such that the robot milking arm was correctly placed relative to the cow's udder when the cow reached the feeding trough. It was found that the cow's length (or more accurately the distance from the cow's mouth to its udder) could vary by up to half a metre. After considerable development the system worked well and it was found that the cows soon learnt that they could be milked whenever *they* wished, rather than the twice-a-day routine used by the humans. As a result, milk yield increased by 10 per cent, and the cows were more contented. A full systems engineering analysis needs to take account of the possibility of the optimum arrangement in any process being different to that envisaged. In this case the cow's preference for milking was beneficial compared to what the human preference was.

7.8.2 The mini-tractor concept and use of satellite images

It is recognized that the sowing of seed in a field should ideally be varied in depth in the soil and in concentration, to suit the immediate soil conditions. The historic techniques of hand sowing or drilling behind a horse or tractor are actually wasteful methods, both in use of the seed and the resulting yield. Use is now being made of satellite imagery to plan seed-sowing depths and concentrations to suit the varying soil conditions in each individual field. There are two approaches as to how to achieve these different requirements. One is to use a single tractor with an intelligent system controlling the seed-feed mechanism; the other which is being considered is to use a number of small, remotely controlled tractors, each sowing particular areas of a field, based on the satellite information about soil depth, composition, fertility, and so on.

The aspect to consider in this example is the improvement in fertility possible by combining totally different technologies in an overall system. It is an approach of looking at farming as a system rather than just as an agricultural skill.

7.9 The effect of accountancy procedures on engineering decisions.

We referred earlier to how local financial requirements can influence systems decisions in an adverse way – see Section 7.3. In an engineering design scenario, use of inappropriate cost information (even when it is the only information available) can have interesting results! Consider the following.

A famous company approached the need to reduce manufacturing costs by instructing all its engineers to query their decisions and actions on a costs basis. As a result, the design office were given the costs of each manufacturing process. A need arose to reduce the heat emitted to the immediate surrounding structure from a powerful gas turbine. The thermodynamics department suggested that a heat shield should be fitted around the engine and that the shield should be plated with a metal having a low emissivity rating. The designer found that the best plating to use was gold, and on looking up the cost figures for gold plating was surprised to find that plating carried no incremental cost increase because all plating processes were charged to overheads. The designer therefore specified gold plating, and to this day the engine carries a gold-plated heat shield – some 3 m × 1 m!

Although this study (which is completely true) does seem ludicrous today, the designers were not used to having to design against cost figures and used the only figures available. The costs of the plating processes were put on overheads because the piece part cost could never be established with any accuracy since the plating baths were run continuously, whether there were any parts in the baths to be plated or not. The cost figures therefore made sense in an overall accountancy analysis but were completely useless for making engineering decisions.

7.10 New technologies associated with systems engineering

In considering case studies, it is useful to consider the new technologies involved, or perhaps where new technologies might help. The following list is not exhaustive but is a useful mental guide to have in front of you.

1. Increased system integration leading to embedded decisions.
2. Increased system integration necessitating self-learning/self-determining intelligent systems.
3. Movement away from historical presumptions in technology solutions.
4. Wide systems implications (away from purely engineering considerations).
5. Through-life implications.
6. Design features versus customer benefits.
7. Additional design to shield user from complexity.

7.11 Examples of technology mix in industrial sectors

7.11.1 Aerospace

(a) Integration of all aircraft and engine control systems [New Techs 1,

2, 3, 4, 5, 6, 7], including fly by wire [New Techs. 1, 4, 5, 6, 7]

(b) Scheduling of air cargo services [New Techs 2, 3, 4, 5, 6]

(c) Air traffic control [New Techs 1, 4, 7]

(d) Active control of the blended wing aircraft concept with the 'all electric' gas turbine [New Techs 1, 2, 3, 4, 6, 7]

7.11.2 Agriculture

(a) Control of tractor implements [New Techs 1, 3, 4, 5, 6, 7]

(b) Automatic milking of cows [New Techs 1, 2, 5, 6, 7]

(c) Multi-mini-tractor concept [New Techs 1, 3, 4, 5, 6, 7]

(d) Farming as a system [New Techs 2, 3, 4, 5, 6, 7]

7.11.3 Automobile

(a) Engine management [New Techs 1, 3, 5, 6]

(b) ABS [New Techs 1, 3, 4, 5]

(c) Traction control [New Techs 1, 3, 4, 5]

(d) Variable valve timing [New Techs 1, 3, 5]

(e) Occupant safety systems [New Techs 1, 4, 5, 6]

(f) Electric brakes [New Techs 1, 3, 4, 5, 6]

(g) F1 versus passenger car [New Techs 1, 3, 7 versus 1, 3, 4, 5, 6, 7]

(h) F1 technology [New Techs 1, 3, 7]

7.11.4 Gas turbines and rotating machines

(a) Intelligent control of blade position and geometry [New Techs 2, 3, 5, 7]

(b) Active magnetic bearings [New Techs 1, 3, 5]

(c) Modular engines [New Techs 3, 4, 5, 7]

(d) The all-electric/oilless gas turbine [New Techs 1, 3, 4, 5, 7]

7.11.5 Manufacturing

(a) Constant-quality manufacture [New Techs 2, 3, 4, 5, 6, 7]

(b) Agile and reconfigurable machines [New Techs 1, 4, 5, 7]

(c) Mass customization [New Techs 2, 3, 4, 5, 7]

7.11.6 Medical engineering

(a) Intelligent limbs [New Techs 1, 5, 6, 7]

(b) Operations by robots [New Techs 2, 3, 4, 5, 7]

(c) Micro-miniature robotics for internal operations [New Techs 2, 3, 4, 5, 6, 7]

7.11.7 Modularity

(a) The Tornado engine [New Techs 1, 4, 5]

(b) Interchangeable systems [New Techs 1, 2, 3, 4, 5, 7]

7.11.8 Process industries

(a) The wrapping wachine [New Techs 1, 4, 5, 6]
(b) Flexible manufacturing lines [New Techs 2, 4, 5, 6, 7]
(c) Buffer store technologies [New Techs 3, 4, 5, 6]

7.11.9 Transportation

(a) Integrated systems [New Techs 2, 3, 4, 5, 7]
(b) Feeder systems [New Techs 1, 4, 5]

7.11.10 Unsuccessful systems

(a) The Passport Office [New Techs 1, 3, 4]
(b) The London Ambulance Service [New Techs 1, 3, 4]

The reader might like to make their own list based on this breakdown; it can also be used as a useful *aide-memoire* on any systems engineering problem.

Chapter 8

Final Remarks

Engineering is all about applying scientific principles to the design and manufacture/implementation of useful items and/or solutions to real-world problems. It is therefore intimately linked with the demands and expectations of the real-world users and customers and with the technological opportunities available at any point in time. This environment has changed and continues to do so, dramatically, and engineering must evolve to meet this challenge. The balance between market pull and technology push has always changed, but the rate of change of this cycle is accelerating. For example, mass customization (bespoke production) demands great agility in manufacturing and hence different engineering design characteristics than the efficiency-driven mass production of the past.

Not only is the environment for engineering design more complex, so too are the technologies which are available. In particular, developments in information technology, software, sensors, and actuation enable novel architectures and intelligent control and so fundamentally change the relationships between individual enabling technologies, with design focused much more on highly integrated systems. This integration has profound effects on the corresponding relationships between the historical engineering disciplines. It demands a paradigm shift in the engineering profession to fully achieve the innovative and joined-up approach necessary to tackle the technology opportunities and the market expectations.

This change should not be thought of as a natural continuation of the trends of the last few centuries – it demands a genuine step-change in the thinking behind engineering design. Historically, while there have been major developments in individual technologies, they have been exploited within existing, well-understood design architectures that themselves evolved in a long-term, tried-and-tested manner. Note, for example, the 'architectural' similarities between the early horseless carriage and its horse-drawn predecessor. Today, we have the capability and enabling technologies to design architectures that are totally novel and hence not subject to the same evolutionary path and confidence. These new opportunities in design are necessary to achieve the more complex real-world expectations; staying with traditional approaches to engineering design is no longer a viable option.

Modern engineering design must be based around a multi-disciplinary systems engineering approach, and this must be market-focused as well as technology-focused. Systems engineering design forms the bridge between market expectations and technology opportunities.

In this book we have explored both of these aspects of change:

- The market/enterprise environment
- The technological enablers.

The many case studies illustrate the profound nature of their implications on engineering design.

These challenges actually make engineering, as a profession, even more exciting, relevant, and wide-ranging than ever. Engineering design, based on multi-disciplinary, integrated systems engineering approaches, is the hub that enables the aspirations demanded of us to be realized.

Index

Acceleration profiles 2
Actuation 5, 69
Actuator 5, 25–27, 62, 71
  medical 5
  miniaturized medical 5
Agents 75
  autonomous 76
  communities of 76
  intelligent 76, 78
  simple autonomous 77
Agility 7
Artificial Intelligence (AI) 11, 25, 71, 72–74, 78
  embedded 83
Automated recognition 79
Automatic control of machines 17
Automatic lawn mower 74
Autonomous agent 75, 76
  simple 77
Autonomous Intelligent Agents (AIAs) 75,77

Behaviour, emergent 75, 77, 78
Bespoke manufacture 6
Bespoke tailoring 8
Browser 75
Brushed DC motor 27
Brushed permanent-magnet 27
Brushless AC motor 27
Brushless DC motor 27
Brute force 80
Bus network 102

Cams 3, 27
Case studies 95
Central heating control 74
Chain drives 3
Cheque reading 74
Chess 72
Classifying 83
Closed-loop control 67
Cognition 69
Communication 71
  technologies 5
Communities of agents 76
Complex mechanical systems 3
Compliance 100
  couplings 100
Component interference patterns 63
Computational Fluid Dynamics (CFD) 60, 63
Computer Aided Design (CAD) 59, 60
  systems 61
Computer-controlled electric drives 15
Computer-controlled linear motors 32
Concepts (evaluation of) 61
Conceptual design 59, 61
  phase 61
Concurrent engineering 65
Constant speed 17
  drive 27
  electric drive 18

machines 18
Constraints analyser 89
Constraints analysis 89
Continuing professional development 49
Control 67, 68
closed-loop 67
engineering 67, 68
intelligence 67
Controlled drive 27
motors 27

Decision making 74, 82
Design:
conceptual 59, 61
departments 60
detail 59
embodiment 59
multi-discipline 2
office practices 59
practice 60
process 59
strategic 10, 61–63
strategy 61
techniques 10
traditional 2
Detail design 59
Deterministic methods 80
Diagnosis 82
Displacement profiles 2, 27
Distributed intelligence 75
Drawing skills 60
Drawing standards 60
Drives, controlled 27
Dynamic network 101
Dynamic scheduling 103

Electric drive:
(constant speed) 18
software controlled 5
Electrical interface 4
Electronic interfaces 21
Embedded artificial intelligence 83
Embedded computing 71
Embedded intelligence 67
Embodiment design 59
Emergent behaviour 75–78
Emergent properties 76, 77
Engineering design 36, 54, 59
Engineering education 49
Enterprise expectations 56
Evaluation of concepts 61
Expert systems 25, 82

Feed-forward circuit 68
Feed-forward methods 67
Finite element analysis (FEA) 60, 63
Finite element method 63
Flexibility 6
Formalized (system) engineering design 54
Full-life implications 21
Fuzzy control 74, 82
Fuzzy logic 25, 74, 82, 83

Gears 3, 27
Genetic algorithms 74, 81
Geneva mechanisms 3, 27
Goods distribution system 101

Heuristic methods 80
Heuristic searching 80
Hill climbing 81
Human controller 18

Idea sourcing 85
Image capture 74, 79
Image interpretation 74, 79
Information Technology (IT) 5, 13
Intelligence:
control 67
distributed 75
embedded 67
Intelligent agents 76, 78
autonomous 75
Intelligent behaviour 25
Intelligent control 2, 3, 19, 21, 26
systems 18

Intelligent Geometry Compressor 77
Intelligent leg joint prosthesis 75
Intelligent machine 69
Interfaces 5
 electronic 21
International Council on Systems Engineering (INCOSE) 38
IT communications 9
IT techniques 9

Just-In-Time (JIT) techniques 103

Knowledge-based systems 82

Lawn mower, automatic 74
Learning 74, 77
Life cycle 61
Linear motors 26
 computer-controlled 32
Linkages 3, 27
Local bus network 102

Machine vision 74, 79
Machines:
 automatic control of 17
 constant speed 18
Management attitudes 7
Management Information System (MIS) 25, 26
Management structures 7
Manufacture, bespoke 6
Manufacturing Resource Planning systems (MRPII) 76
Manufacturing:
 supply chains 77
 systems 103
Mass customization 6–10
Mass distribution 8
Mass marketing 8
Mass media 8
Mass production 8
Material sciences 14
Mechanical design simulations 63
Mechanical governing 18
Mechanical interfaces 4, 15
Mechanical sub-systems 23, 100
Mechanical systems, complex 3
Mechanical transmission systems 3
Mechatronic systems 5
Mechatronics 24, 27–29
Medical actuators 5
 miniaturized 5
Metallurgy 14
Microcomputer-based systems 21
Military thinking 9
Miniaturization 5
Miniaturized medical actuators 5
Modularity 7
Molins OASIS 24
Motor:
 brushed DC 27
 brushless AC 27
 brushless DC 27
 controlled drive 27
 reluctance 28
 stepper 28
Multi-Agent Systems (MAS) 25, 102
Multi-agent technology 25
Multi-discipline design 2
Multi-discipline systems approach 3, 10

Neural networks 74, 78, 79
Noughts and crosses 72

Pattern matching 74, 78, 83
Perception 69
Perception–Cognition–Actuation model 69
Phase synchronized drives 28
PLC 98
Poker 72
Position profile 2, 3
Pragmatic Principles of Systems Engineering 38
Probabilistic methods 80, 81
Programmable Logic Controller (PLC) 24, 72
Project milestones 60
Property, emergent 76, 77

Proportional, Integral, Derivative (PID) controller 24
Prosthesis, intelligent leg joint 75
Prototyping techniques 63

Rapid prototyping techniques 60
Reactive behaviour 69
Recognition 74
Reconfigurable systems 10
Reliability 32
Reluctance motors 28
Robot football 72
Rule-based systems 74, 82

Search space 80
Searching 74, 75, 83
  algorithms 74
  techniques 79
Self-correcting machine 24
Self-learning control 21
Self-learning systems 26
Sensing 69
Sensor 5, 24, 25, 26, 62, 69, 71
  fusion 70
Signal conditioning 71
Simple autonomous agents 77
Simple systems 4, 5
  technologies 4
Simulated annealing 81
Simulation 60, 63–65
Simulation techniques 63
Single-discipline solutions 3
Software controlled electric drives 5
Speed control 17
Standard subset routines 60
Standardization 10
Statistical Process Control (SPC) 24, 26
Steering sub-system 19
Stepper motors 27, 28
Stovepipe sectors of thought 9
Strategic design 10, 61-63
Strategic development 87
Strategic management 87
Sub-system 4, 5, 10
  integration 7, 21
  steering 19
Supply chains, manufacturing 77
Swarms 77
System integration 11
System lifetime issues 10
Systems:
  approach 46
    multi-discipline 3
    behaviour 65
  engineering 2, 3, 7–11, 32, 59, 61, 95, 102
    approach 32, 38, 46, 95
  integration 9, 10, 60
  knowledge-based 82
  rule-based 82

Technical specifications 62
Technology pull 7, 8
Technology transfer 90
Telechirs 72
Timetabling system 102
Traditional design 2
Transducers 24, 62
Tree searching 80

Unified Modelling Language (UML) 40

Valve driving mechanism 30
Valve gear 30
Velocity profiles 2, 3, 27
Virtual Design Studio 60
Virtual meeting 60
Virtual reality process 60
Vision 74
Vision system 74, 75, 78

Web browsers 75
Whole life costs 32
Work station 97
World Wide Web (WWW) 75
Wound-field-type motors 27
Wrapping machine 95